Emissions Trading Design

NEW HORIZONS IN ENVIRONMENTAL AND ENERGY LAW

Series Editors: Kurt Deketelaere, *Professor of Law, University of Leuven, Belgium and University of Dundee, Scotland* and Zen Makuch, *Reader in Law, Barrister, Imperial College, London, UK*

Environmental law – including the pressing considerations of energy law and climate change – is an increasingly important area of legal research and practice. Given the growing interdependence of global society and the significant steps being made towards environmental protection and energy efficiency, there are few people untouched by environmental and energy lawmaking processes.

At the same time, environmental and energy law is at a crossroads. The command and control methodology that evolved in the 1960s and 1970s for air, land and water protection may have reached the limit of its environmental protection achievements. New life needs to be injected into our environmental protection regimes – perhaps through the concept of sustainability in its environmental, economic and social forms. The same goes for energy policy and law, where liberalisation, environmental protection and security of supply are at the centre of attention. This important series seeks to press forward the boundaries of environmental and energy law through innovative research into environmental and energy law, doctrine and case law. Adopting a wide interpretation of environmental and energy law, it includes contributions from both leading and emerging international scholars.

Titles in the series include:
Conservation on the High Seas
Harmonizing International Regimes for the Sustainable Use of Living Resources
Simone Borg

Global Environmental Governance
Law and Regulation for the 21st Century
Louis Kotzé

Climate Law in EU Member States
Towards National Legislation for Climate Protection
Edited by Marjan Peeters, Mark Stallworthy and Javier de Cendra de Larragán

International Law and Freshwater
The Multiple Challenges
Edited by Laurence Boisson de Chazournes, Christina Leb and Mara Tignino

Genetic Resources, Equity and International Law
Camena Guneratne

Energy for the 21st Century
Opportunities and Challenges for Liquefied Natural Gas (LNG)
Susan L. Sakmar

Emissions Trading Design
A Critical Overview
Stefan E. Weishaar

EU Environmental Legislation
Legal Perspectives on Regulatory Strategies
Edited by Marjan Peeters and Rosa Uylenburg

Polar Oceans Governance in an Era of Environmental Change
Edited by Tim Stephens and David L. VanderZwaag

The Fragmentation of Global Climate Governance
Consequences and Management of Regime Interactions
Harro van Asselt

Emissions Trading Design

A Critical Overview

Stefan E. Weishaar

Associate Professor of Law and Economics, University of Groningen, The Netherlands

NEW HORIZONS IN ENVIRONMENTAL AND ENERGY LAW

Edward Elgar

Cheltenham, UK • Northampton, MA, USA

Published by
Edward Elgar Publishing Limited
The Lypiatts
15 Lansdown Road
Cheltenham
Glos GL50 2JA
UK

Edward Elgar Publishing, Inc.
William Pratt House
9 Dewey Court
Northampton
Massachusetts 01060
USA

A catalogue record for this book
is available from the British Library

Library of Congress Control Number: 2013947144

This book is available electronically in the ElgarOnline.com Law Subject Collection, E-ISBN 978 1 78195 222 1

MIX
Paper from
responsible sources
FSC
www.fsc.org
FSC® C013056

ISBN 978 1 78195 221 4

Typeset by Columns Design XML Ltd, Reading
Printed and bound in Great Britain by T.J. International Ltd, Padstow

Edwin Woerdman, Associate Professor of Law and Economics, University of Groningen, The Netherlands has written large parts of Chapter 5 and the passage on personal carbon trading. His authorship of these sections is gratefully acknowledged. He has kindly permitted the use of his work in this book.

Contents

Abbreviations

AAU	assigned amount unit
ACCC	Australian Competition and Consumer Commission
ACCU	Australian carbon credit unit
ACU	Australian carbon unit
CARB	California Air Resources Board
CCS	carbon capture and storage
CDM	clean development mechanism
CER	certified emission reduction
CFI	Carbon Farming Initiative
CH_4	methane
CITL	Community Independent Transaction Log
CO_2	carbon dioxide
CPM	Carbon Pricing Mechanism (Australia)
EITE	energy-intensive trade-exposed
EPA	Environment Protection Agency (US)
ERC	Emission Reduction Credit
ERU	emission reduction unit
ETS	emissions trading system, emissions trading scheme
EU	European Union
EU ETS	European Union Emissions Trading Scheme
EUA	European Union allowance
EUTL	European Union Transaction Log
GATS	General Agreement on Trade in Services
GATT	General Agreement on Tariffs and Trade
GDP	gross domestic product
GHG	greenhouse gas
GUO	Global Ultimate Owner
HFCs	hydrofluorocarbons

ICAO	International Civil Aviation Organization
IEU	international emissions unit (Australia)
IPCC	Intergovernmental Panel on Climate Change
ITL	International Transaction Log
JI	joint implementation
LULUCF	land use, land-use change and forestry
MAD	Market Abuse Directive
MAR	Market Abuse Regulation
MGGRA	Midwestern Greenhouse Gas Reduction Accord
MiFID	Market in Financial Instruments Directive
MoU	Memorandum of Understanding
MRV	monitoring, reporting and verification
MTIC	missing trader intra-community
MW	megawatts
N_2O	nitrous oxide
NAP	National Allocation Plan
NF_3	nitrogen trifluoride
NGO	non-governmental organization
NOx	mono-nitrogen oxides
NZ ETS	New Zealand Emissions Trading Scheme
NZU	New Zealand unit
OECD	Organisation for Economic Co-operation and Development
PFCs	perfluorocarbons
PSR	performance standard rate
RGGI	Regional Greenhouse Gas Initiative
RMB	renminbi
RMU	removal unit
SCM	Subsidies and Countervailing Measures
SF_6	sulphur hexafluoride
SMEs	small and medium-sized enterprises
SO_2	sulphur dioxide
SOE	state owned enterprise
TEU	Treaty on European Union
TFEU	Treaty on the Functioning of the European Union
TMG	Tokyo Metropolitan Government

UN	United Nations
UNFCCC	United Nations Framework Convention on Climate Change
US	United States
VAT	value added tax
WCI	Western Climate Initiative
WTO	World Trade Organization

Table of cases

Australia

European Court of Human Rights

European Union

The Netherlands

New Zealand

United Kingdom

United States

1. Introduction

1. RELEVANCE OF EMISSIONS TRADING

Not that long ago, hardly anyone would have thought that greenhouse gas emissions could, and would, be an object of trade and that such a process would help to reduce these emissions. Despite fierce criticisms directed at emissions trading, the system is in full operation in the context of climate change, and it is spreading around the globe. Emissions trading was first employed in the United States (US) and was then embraced by the United Nations Framework Convention on Climate Change (UNFCCC). Emissions trading systems were then created in a few countries in Europe before being adopted by the European Union (EU), and have recently also been introduced in Oceania and Asia.

Seen from a proliferation perspective, emissions trading is a great success, having been created and crafted from nothing more than the virtual world of economic and legal ideas. Emissions trading systems can, of course, be used in other contexts to reduce other gases – such as sulphur dioxide (SO_2) (which causes acid rain) or mono-nitrogen oxides (NOx) (which cause smog and acid rain) – but in this book we concentrate on climate change as its most prominent area of application at present. Gases responsible for climate change are carbon dioxide (CO_2), methane (CH_4), nitrous oxide (N_2O), sulphur hexafluoride (SF_6), and two groups of gases, hydrofluorocarbons (HFCs) and perfluorocarbons (PFCs).

Emissions trading is one of several policy instruments that policy makers can use to achieve environmental objectives. Emissions trading is a 'market based' instrument that allows polluters (installations that emit greenhouse gases, generally referred to as emitters) to decide for themselves who should undertake greenhouse gas abatement activities. Under an emissions trading scheme, emissions that formerly involved no cost now have a price. The emitter with the lowest abatement costs will sell its abatement efforts in the form of emission allowances to polluters who have higher abatement costs. Since emitters with the lowest abatement costs will undertake the abatement efforts, it is a cost-effective means of reducing emissions. Because emissions trading attributes a larger role to entrepreneurial decision making than does traditional ‘command and control’ regulation – while avoiding the negative stigma of taxation – it is expanding rapidly.

2. HISTORY OF EMISSIONS TRADING

Emissions trading was first proposed in 1968 by the North American scientist, J.H. Dales, in his book *Pollution, Property and Prices: An Essay in Policy-Making and Economics*. In 1975 the US Environment Protection Agency (EPA) began to experiment with forms of emissions trading to control air pollution. Since then the instrument has been used in various other US programmes, for instance, to reduce ozone-depleting substances under the Montreal Protocol of 1989.[1] But probably the most well-known successful early example of emissions trading is the US sulphur dioxide (SO_2) emissions trading program for electricity producers which started in 1995.

From there, discussions on emissions trading went to the United Nations (UN). In December 1997 the Kyoto Protocol was signed by which industrialized countries committed themselves to reduce six greenhouse gases during the period 2008–12, on average by 5 per cent per year, to limit greenhouse gas emissions below 1990 levels.[2] These commitments to reduce greenhouse gas emissions were legally binding.

States that accepted a binding emissions reduction target were referred to as ‘Annex I countries’. The Kyoto Protocol could enter into force on 16 February 2005, when two conditions were met. At least 55 states to the UNFCCC (which incorporated the Annex I countries) had to ratify the Protocol and these countries had to account for at least 55 per cent of the total carbon dioxide emissions for 1990 of the parties included in Annex I. The first of the two conditions, the ‘55 parties’ clause, was met in May 2002 when Iceland ratified the Protocol. The ratification by Russia on 18 November 2004 satisfied the ‘55 per cent’ clause and (after 90 days) brought the Convention into force.

The Kyoto Protocol also introduced two project based mechanisms: the Clean Development Mechanism (CDM) and the Joint Implementation mechanism. The CDM[3] allows parties listed in Annex I of the Kyoto

---

[1] Montreal Protocol on Substances that Deplete the Ozone Layer, Montreal (Canada), 16 September 1987, in force 1 Jan. 1989, available at: http://ozone.unep.org/new_site/en/montreal_protocol.php.

[2] Kyoto Protocol, Art. 17.

[3] Ibid., Art. 12.

Protocol to finance emission reduction projects in countries which are not listed in Annex I. Annex I parties are awarded certified emission reductions (CERs) for doing so. CDM projects assist non-Annex I parties in pursuing sustainable development policies and in contributing to the ultimate objective of the Convention, and assist Annex I parties in meeting their emissions targets. The Joint Implementation (JI) mechanism, by contrast, allows parties listed in Annex I of the Kyoto Protocol to receive emission reduction units (ERUs) for co-financing projects that reduce net emissions in an Annex I country. Parties are awarded ERUs from JI projects.[4] Excessive use of these flexibility instruments is intended to be curtailed by the supplementarity requirement of the Marrakech Accords.[5] Supplementarity is defined as a mechanism the use of which is to be 'supplemental to domestic action and that domestic action shall thus constitute a significant element of the effort made by each Party included in Annex I to meet its quantified emission limitation and reduction commitments under Article 3, paragraph 1'.[6]

Even though the Kyoto Protocol was not embraced by the United States – which, at that time, was the largest emitter of greenhouse gases – in the EU emissions trading was still seen as an appealing policy tool. After Denmark[7] and the United Kingdom[8] started to adopt emissions trading, the EU decided, in 2003, to implement a similar system. Since EU law supersedes national law in the area of climate change and the EU dominates this field of law, these national schemes had to comply with the new requirements. The European Emissions Trading Scheme (the EU ETS) was, to a large extent, inspired by the US SO_2 trading system. The EU ETS started in 2005 as a multi-jurisdictional attempt to reduce CO_2 emissions from four broad sectors:

- energy (electric power, oil refineries and related industries);
- the production and processing of ferrous metals (iron and steel);
- minerals (cement, glass, ceramics); and
- pulp and paper.[9]

4 For an economic review of CDMs and JIs see Brander (2003). The linking between the EU ETS and flexibility mechanisms is governed by Directive 2004/101/EC. See De Cendra de Larragán (2006).

5 Marrakech Accords (2001). These accords describe the rules for meeting the targets of the Kyoto Protocol.

6 See Marrakech Accords (2001) p. 51.

7 See Lauge Pedersen (2000) and Zwingmann (2007) pp. 113ff.

8 See Zwingmann (2007), pp. 117ff.

9 Directive 2003/87/EC, Annex I.

The EU ETS now covers 31 countries (all 28 EU Member States and the three European Economic Area countries: Iceland, Liechtenstein and Norway) and also includes the aviation sector. It sets a cap on approximately 40 per cent of the European Union's greenhouse gas emissions.

The EU ETS was implemented over several trading phases: Phase 1 ran from 2005 to 2007, Phase 2 from 2008 to 2012, and Phase 3 runs from 2013 to 2020. The first trading period was described as a 'learning by doing' phase. The system suffered from over-allocation, which – when it became public knowledge – led to a crash in the emission allowance price from €28 to ultimately €0.10 in 2007. Over-allocation is partly attributable to data problems and to over-optimistic company growth expectations. An attempt can be made to explain it by successful lobbying and the understanding of Member States that a more generous allocation to domestic installations would help them to retain competitiveness in the international market.

Moreover, politically unacceptable 'windfall profits' were created. They stemmed from the ability of the electricity sector to pass on cost increases to customers.[10] The learning process still continued beyond the first trading phase. Despite the amendments made to the National Allocation Plans of the EU Member States by the European Commission, an oversupply situation arose again in the second trading phase (2008–12) for which there were no adequate systemic safeguards.[11] This oversupply is still a problem in the third trading phase (2013–20) and, among other factors, is attributable to the economic downturn in 2009. To a lesser extent it is also attributable to the increased use of Kyoto offsets and to the sale of new entry reserves that came onto the market in 2012. As a result of this situation, the EU allowance prices dropped in Phase 2 and still remain at levels that are too low to incentivize investment in greenhouse gas emission abatement and innovation. Currently the price of a European Union allowance (EUA) is €3.50.[12] Technological innovation is one of the (sub) policy objectives listed in the EU ETS

---

[10] The electricity sector, in particular, faces 'inelastic demand'. This means that as prices increase, consumers only marginally change their consumption behaviour. Energy companies were thus able to pass on large parts of their production cost increases to consumers even though they had received allowances for free. This was perceived as 'unfair' by consumers and politicians alike. See Woerdman, E., O. Couwenberg and A. Nentjes (2009).

[11] See Peeters and Weishaar (2009) pp. 94 and 95.

[12] The spot market price for EUAs at the Leipzig Energy Exchange on 28 May 2013, available at: http://www. eex.com.

Directive.[13] It should be mentioned, however, that these problems do not undermine the system's effectiveness to achieve emission reductions at low costs.

Because of their cost-saving potential for society, emissions trading systems remain attractive to policy makers around the world. Other systems have emerged in Europe (Switzerland, for example), and also in North America (the Regional Greenhouse Gas Initiative (RGGI) and the Western Climate Initiative (WCI)). Emissions trading systems are operating in New Zealand and Australia, and also in Japan (for example, the scheme set up by the Tokyo Metropolitan Government (TMG)) and in Kazakhstan. Six pilot schemes are currently been set up in China, and South Korea is also developing a scheme. India Thailand, Vietnam and Mexico are also reported to be considering the introduction of a system of emissions trading.

Given the current developments and the spread of emissions trading, it is likely, although not certain, that more countries will follow in adopting emissions trading as a cost-effective instrument for environmental policy. It is also possible, but again not certain, that countries will link their emissions trading schemes in order to create markets of higher density with more efficient abatement opportunities, even though this would necessitate overcoming various political, legal and economic barriers. Given the continuing spread around the globe, the relevance of emissions trading, as well as of the various possibilities for design choices, gains more importance.

3. ASPECTS OF EMISSIONS TRADING DESIGN

Emissions trading is one of several – and is probably one of the most important and efficient – policy tools that can be chosen to address climate change. As is the case with any tool, it can be used to achieve a goal: a hammer is used to hammer; a greenhouse gas emissions trading system is used to reduce greenhouse gases.

A policy maker wishing to address climate change will look into an instrument tool box and determine which instrument will work best for the desired objective. The instrument options (command-and-control regulation, liability rules, taxation and emissions trading) and their merits are discussed further in this book. As important as the tool itself are its particular design features. These design features crucially depend on

[13] Directive 2009/29/EC, recital 8.

what precisely the tool is intended to do and on the context in which it needs to operate.

Policy makers who consider the adoption of an emissions trading system should therefore consider first the particular context in which the scheme will be operating and decide what the instrument is supposed to achieve. There are many objectives that policy makers would like to see realized, but not everything can be achieved at the same time – and certainly not with every emissions trading design. An example from the EU ETS is illustrative. The system was originally designed to reduce greenhouse gas emissions at the lowest possible cost and, at a later point in time, it was also required to incentivize technological innovation. The system, however, is not designed to deliver high emission allowance prices at times of economic downturn. This inability has given rise to severe criticism of emissions trading in general and the EU ETS in particular, while it is evident that the system is doing precisely what it was designed to do and that it cannot deliver something it was not designed to achieve. Emissions trading designers are therefore well advised to have a clear list of what they would like to achieve before starting work on designing such a system.

4. PURPOSE OF THE BOOK

Emissions trading design is clearly important. The purpose of this book is to provide a critical overview of these design issues. The central questions therefore are: (i) how to design an emissions trading scheme; (ii) what are the potential implementation problems; and (iii) how can they be addressed?

The following sub-questions are addressed:

- What are the main advantages and disadvantages of emissions trading in general?
- What needs to be considered in designing an emissions trading system?
- What are the main implementation problems of emissions trading?
- How should those implementation problems be addressed in an effective, efficient and acceptable way?

In order to address these questions the structure of the book will follow four simple questions relevant for emissions trading designers: why, how, what, and with whom?

Next to the design choices that policy-makers need to make in order to reach their policy objectives, we focus on the main implementation problems of emissions trading. These are the issues surrounding the initial allocation of emission rights, the secondary markets for these rights, compliance with emissions trading rules, and lawsuits resulting from emissions trading schemes. Some of these implementation issues we elucidate by taking the European Union Emissions Trading Scheme (EU ETS) as an example – not because it is the best system that exists, but rather because it offers useful examples of issues that system designers would like to avoid. Also addressed is the timely issue of linking emissions trading schemes. Negotiations to establish unilateral links between Australia and the EU ETS on one side and bilateral links between the EU ETS and the Swiss system are well under way. It would be useful to look at these links-in-the-making and pick up some lessons on design.

Economists like to emphasize that 'the devil is in the detail' – which is true. Details can matter a great deal for the performance of emissions trading schemes. Some of these details will be highlighted and analysed, especially in relation to the EU, which is our home base and primary field of expertise. However, in a book such as this it is impossible to go into every detail of all relevant laws and regulations for the various emissions trading schemes in multiple jurisdictions. This would take away the focus from the most salient issues that our overview intends to provide.

Emissions trading systems are being set up in many jurisdictions around the world, but all policy makers face similar design challenges. In outlining these challenges, this book is unique. There is much scientific literature on emissions trading that deals with very specific questions, but an up-to-date and easily accessible book presenting and critically reflecting on how to design such a system and which implementation challenges will be encountered is missing. Moreover, the current literature typically focuses on the environmental, legal and economic aspects of emissions trading and pays far less attention to practical policy problems, from initial allocation to compliance and lawsuits. Our book addresses this niche by targeting policy makers, managers, consultants, practitioners and graduate students, as well as an audience that needs to familiarize itself with emissions trading design in a comprehensive and time-efficient way, with an approach that takes the multiple policy objectives of ETS designers as a central element.

Following this introductory chapter, Chapter 2 reviews the merits and demerits of emissions trading by comparing these schemes with other policy instruments that could be selected to achieve the goal of reducing greenhouse gas emissions. In this way we focus on the reasons 'why', and in which circumstances emissions trading can be considered an effective policy choice. We pay particular attention to efficiency, environmental effectiveness, acceptability, and present some points for consideration for emissions trading designers.

Chapters 3 and 4 address the question of 'how' emissions trading systems can be designed and the factors that policy makers should bear in mind. We will take a look at a policy maker's goals and the wider framework that must be considered when designing an emissions trading system, highlighting the important trade-offs that policy makers face. Chapter 4 presents emissions trading systems in various countries and shows the richness of design solutions that are employed in practice.

Chapters 5 to 8 address 'what' are the special issues in designing an emissions trading system that can pose challenges for its operation. Chapter 5 addresses the initial allocation of emission allowances. Chapter 6 deals with secondary markets, fraud and oversight issues. Chapter 7 deals with operational issues, which include monitoring, verification and enforcement; it also addresses issues related to emissions trading transaction logs. Chapter 8 reviews the case law that has arisen in the context of climate change and emissions trading systems, and seeks to draw the awareness of designers to the legal issues. This is particularly crucial since any policy instrument will ultimately need to operate within a regulatory, and therefore, legal framework.

Chapter 9 looks at the possibilities of expanding the benefits of an emissions trading system by linking it to similar systems. Since design features may determine the ability to link systems, this chapter addresses the question of 'with whom' to establish a trading system. Chapter 10 summarizes the main findings.

5. ACKNOWLEDGEMENTS

This book has taken shape thanks to the continuous support of the members of the research group on Energy and Sustainability at the Department of Law and Economics of the Law Faculty of Groningen University. Without their tremendous support and encouragement this project could never have been realized. Special thanks go to Oscar Couwenberg for being most enabling and encouraging in the process of writing this book. Special thanks also go to Edwin Woerdman – whose encouragement and constant eye for detail and out-of-the-box thinking has provided an invaluable source of inspiration and critique. His valuable contribution to this book – the passage on personal carbon trading and large parts of Chapter 5 – are gratefully acknowledged. Both Oscar and Edwin also deserve my gratitude for their invaluable comments on the manuscript of this book.

Last, but not least, I wish to thank our PhD researchers who helped in gathering literature, providing ideas and writing some text fragments for various chapters. It is scientifically most encouraging and rewarding to be able to work with them. I want to thank especially Charis van den Berg for contributing to chapters 4 and 8, Suryapratim Roy for his contribution to chapter 8, Fitsum Tiche for contributing to chapters 4 and 9, and Thijs Jong for contributing to the section on the transaction logs.

Research for this book ended in the beginning of the year 2013. More recent developments have therefore not been taken into account. There are, however, three developments that should be mentioned:

1. The back loading proposal of the European Commission has been accepted by the European Parliament on the 3rd of July 2013 and the Council reached an agreement on the 8th of November 2013. As a result, legislation for back loading may be adopted before April 2014.
2. The Chinese emissions trading pilots in Beijing and Shanghai have commenced operations in November 2013.
3. After defeating the Labour Party in the recent Australian federal election, the National Coalition-led government proposed to scrap the CPM. Given that the ruling party lacks a majority in the Senate, it is not yet certain if the proposal will succeed.

2. Emissions trading and alternative instruments[1]

1. INTRODUCTION

In this chapter we compare regulatory approaches to climate change and outline their advantages and disadvantages. Subsequently, we analyse in which situations an emissions trading system can be employed to ensure efficient and effective environmental protection. We pay particular attention to insights to be derived from this analysis that are important for emissions trading designers.

We will not dive into the important question of how to determine the desired level of environmental protection, but a few general remarks on climate change provide a useful background to the analyses. Clearly, industrial production (as a proxy for economic activity) is associated with greenhouse gas emissions. Since it also gives rise to benefits, the level of emissions that maximizes social welfare is positive and not zero. Setting a greenhouse gas emissions target is subject to (scientific) uncertainty and to collective action problems that obstruct the setting of international targets, which would ultimately need to be translated into rule-making for various economic sectors at national level. With regard to (international) target setting it bears mentioning that there are multiple potential goals. The Intergovernmental Panel on Climate Change (IPCC) Fourth Assessment Report (2007), for example, states that if industrialized countries (Kyoto Protocol Annex I parties) want to attain a low stabilization goal of 450 ppm, they should aim at CO_2-equivalent emission reductions of 25–40 per cent in 2020 compared to 1990 levels. If the temperature increase is limited to 2–2.4 degrees, 'then global emissions should, theoretically, be reduced by 60–80 per cent immediately', and if one chooses a target of 400 ppm CO_2-equivalent, global warming could be constrained to less than 2°C with 75 per cent probability.[2] There are also

[1] This chapter constitutes an extended version of a similar treatment contained in Faure and Weishaar (2012).

[2] University of Copenhagen (2009) Synthesis Report, pp. 18 ff.

those who argue that a goal of 350 ppm should be adopted to prevent irreversible climate change damage.[3] Whichever global objective is chosen, it will need to be translated into national actions and targets.

There are various instruments that can be used to reduce greenhouse gas emissions. They can be divided into two groups: 'command and control' type regulation and 'economic incentive-based or market-based' instruments. Command and control regulation is generally understood as the direct regulation of an industry or activity by legislation that prescribes what is permitted and what is not. In contrast to this type of regulation, the economic literature suggests the use of economic incentive-based or market-based instruments. These instruments specify a particular environmental goal, but then leave it to market participants to decide how to reach this goal. These instruments are typically based on financial incentives. Taxes and emissions trading systems are examples of such economic incentive-based or market-based instruments.

The division between command and control regulation and economic instruments is not clear cut because liability rules often rely on financial sanctions and are therefore also 'economic incentive based'. To be effective, both require a particular form of legal design and regulation, as well as enforcement. Because the instruments have different strengths and weaknesses, in the literature much attention is paid to optimal mixes of policy instruments.

This chapter discusses the strengths and weaknesses of command and control regulation as well as economic incentive-based or market-based instruments. Issues such as flexibility, cost-effectiveness and political acceptability are also examined. Such a comparative perspective is useful given the different, and at times seemingly contradictory, approaches to climate change policy design. We show that it is not surprising that there are differing approaches since none of the policy instruments (command and control type regulation, liability rules, taxation and emissions trading) alone may be optimal to provide adequate incentives for greenhouse gas emission control. As is often found in policy practice, a combination may be necessary, which ideally makes optimal use of the advantages of each particular policy instrument. This chapter first addresses command and control type regulation (Section 2), then moves on to examine liability rules (Section 3). Taxes and emissions trading are then reviewed (Sections 4 and 5). Concluding remarks highlight the main findings with regard to emissions trading design.

---

3 Hansen (2009).

2. COMMAND AND CONTROL VERSUS ECONOMIC INCENTIVE-BASED OR MARKET-BASED INSTRUMENTS

Traditionally, environmental law largely consisted of a 'command and control' type of regulation. Economists increasingly pointed to several disadvantages of imposing pollution standards via government regulation.[4] In recent years policy makers have picked up Pigou's message of the 1920s that taxation could be used to internalize externalities. Moreover, since the 1960s, Dales has pointed out that social welfare could be increased by trading in pollution rights.[5] The literature refers to these as economic incentive-based or market-based instruments. Such instruments have been advanced for a number of years by policy analysts as a reaction to command and control regulation. Using the market instead of relying on regulation provides flexibility and has a modern connotation; hence, politicians increasingly argue that environmental policy should be market-oriented instead of merely relying on top-down regulation by government.

Terminology can, however, be misleading. One could argue that all legal and policy instruments could be considered 'economic' as long as they offer appropriate incentives to polluters for an efficient reduction of the externality caused by the pollution. Thus liability rules are also 'economic' in the sense that liability provides incentives to polluters to prevent environmental harm. Nevertheless, liability rules are traditionally not included within the category of 'economic' instruments – a concept that typically refers to instruments in accordance with which only the environmental targets are defined, and it is left to the polluters to find the optimal way for reaching these targets.

2.1 Weaknesses of Command and Control

1. Traditional command and control systems are based on permits and licensing. Since permits set emissions standards without limiting the actual quantity of pollution according to plant activity levels, such regulatory approaches have been criticized since the 1970s for focusing on controlling emissions while disregarding the effect of the emissions on environmental quality. Despite the presence of stringent controls, environmental quality seriously degraded. The

---

4 Faure, Peeters and Wibisana (2006).

5 Dales (1968).

instrument's focus should be on target standards or on environmental quality standards to take due account of activity levels. This criticism does not invalidate *per se* a command and control approach since the instrument could also be used within a system that focuses largely on target standards. Other criticisms of command and control are presented below.[6]

2. Command and control requires optimal standards and optimal enforcement, which involves high information and enforcement costs. If enforcement is less than optimal, profits can be earned by creating more pollution.[7] Although enforcement and information costs could be reduced by relying on industry information, the private interest theory of regulation suggests that industry can benefit from distorting information.
3. Compliance with regulatory standards (such as a permit) does not provide additional incentives for reducing pollution or investing in environmentally friendly technology beyond what is required by law.
4. Polluters bear the cost of complying with regulatory standards. They may, however, not be required to pay for any residual damage associated with pollution created while they are complying with the standard. Adhering to the law may shift the liability for environmental damage to the government.
5. Command and control cannot ensure the equalization of marginal pollution costs among different polluters. Given prohibitively high information and regulatory costs, administrative agencies will not be able to determine the efficient level of emissions for each polluter and thus demand comparable reductions by polluters. This is inefficient, and it increases social welfare losses if the possibilities of pollution abatement differ.[8] Furthermore, polluters which have the opportunity to abate pollution above the regulatory standard have no financial incentive to do so.

This shows that there are considerable disadvantages with the traditional command and control approach: regulatory standards are often too general and not sufficiently flexible and differentiated. An optimal environmental policy needs adaptable instruments (hence the reference to 'market-based') that, on the one hand, provide greater flexibility (taking into account each polluter's individual opportunity for optimal pollution

---

6 For a summary of these criticisms see also Faure and Ubachs (2003).
7 See Tietenberg (2000).
8 Rosen (1999).

abatement) and, on the other hand, provide optimal incentives towards environmental technological innovation rather than simply striving for compliance with a regulatory standard. Having addressed the disadvantages of the traditional command and control approach, we better understand why environmental policy analysts are increasingly interested in the so-called economic instruments.

3. ENVIRONMENTAL LIABILITY AS AN INCENTIVE-BASED INSTRUMENT

It is not only command and control approaches that can be used to address environmental pollution. Legal rules governing the attribution of liability to the parties involved in an accident situation can be used to induce polluters to emit less. In the context of climate change law, the liability of entities for global warming is therefore an important area of research.[9] The underlying economic idea is that liability will provide actors with incentives for careful behaviour. In our case, they might serve as an incentive for entities to emit less greenhouse gas pollution. Economists stress the deterrent function of tort law and thereby take an *ex ante* perspective.

From an economic perspective, the main goal of liability rules is the minimization of 'primary accident costs':[10] the costs of accident avoidance and the damage expected to result from an accident. From a social point of view, investment in accident avoidance represent the costs associated with accidents. Clear examples include not only investment in safety controls, but also less visible costs such as exercising special care in dangerous operations. It is further differentiated between so-called unilateral accidents, in which it is only the care exercised by one of the parties (the injurer) that influences the risk of accident, and bilateral accidents, in which the behaviour of both parties influences the risk.[11] In a bilateral accident situation, therefore, the goal of accident law should be to define incentives to minimize the total cost of the care exercised by the potential injurer and the potential victim and the damage that is expected to occur in the event of an accident.

Economists use classic cost–benefit analysis to determine the level of care that will minimize the social cost of accidents. Such 'optimal or

---

[9] The diversity of topics considered in this area is well presented by the articles contained in Faure and Peeters (2011).

[10] Calabresi (1961); Calabresi (1970).

[11] Shavell (1987).

efficient care levels'[12] are found at the point at which the marginal cost of exercising care equals the marginal benefits of accident reduction.[13] This is true for a risk-neutral setting. Risk-averse actors may prefer a higher investment in care.

Having discussed the regulatory function of liability rules, the relevant question is then which liability rule will provide appropriate incentives to optimally prevent environmental damage. In general, there are two types of classic legal liability rule: the fault-based or negligence rule and the strict liability rule. Under the former, the injuring party compensates the victim only when his behaviour falls short of a certain level of due care as determined by the court. Strict liability, on the other hand, places upon the injuring party a duty to compensate the victim irrespective of the former party's behaviour, provided there is a causal relationship between the activity and the damage.

Economic literature holds that if a negligence rule is adopted, the injuring party will take optimal care, provided the level of due care required by the legal system is equal to the optimal care level.[14] Since an individual will minimize costs by complying with this due care standard, the negligence rule will lead to an efficient outcome.

A rule of strict liability will lead to the optimal level of care in the situation in which only one party is able to influence the risk of accident because the injuring party will bear all of the social cost of accidents, namely his own cost in exercising care and the expected damage.[15] A potential injurer will minimize his total expected accident costs by comparing the marginal cost of exercising care with the marginal benefits of accident reduction.

Considering only the influence of the injuring party's standard of care in an accident situation, both negligence and strict liability can provide economically appropriate incentives for taking optimal care. However, there are important nuances. Administrative and legal costs for strict liability seem to be higher since legal proceedings will usually follow given that the injuring party will always be liable to compensate. On the other hand, the negligence rule has higher information costs for the judge, who will have to determine the marginal costs and marginal benefits of exercising care in order to determine if the injuring party has taken 'due care'.[16]

---

12 Landes and Posner (1984); Polinsky (1983).
13 Shavell (1987).
14 Shavell (1987); Calabresi (1975).
15 Polinsky (1983); Shavell (1987).
16 Brown (1973); Calabresi (1975); Shavell (1987).

In a bilateral accident situation, a contributory or comparative negligence defence needs to be added to a rule of strict liability to give (potential) victims an incentive to take optimal care. In such a situation, a negligence rule will induce victims to take due care. A knowledgeable victim will assume that the potential injuring party will take due care to avoid liability; so the victim will avoid accidents in order not to be left with his loss. Further refinement can be found in the literature, which focuses, inter alia, on the activity levels of dangerous operations.[17]

Environmental pollution can, in most cases, be considered as a unilateral accident situation. Since the victim is unable to influence the accident risk, strict liability would appear to be the best solution to provide the potential polluter with incentives for adopting optimal activity levels and for exercising efficient care.[18]

In cases in which parties other than the polluter could influence the risk of environmental degradation, liability rules should provide incentives for those parties to take appropriate preventive measures. In cases where the third party is also a victim, environmental pollution would not constitute a unilateral risk, but a bilateral risk. A strict liability rule is justified to give the party with the most influence on the risk (the polluter) the incentive to take preventive measures. In bilateral risk situations a defence should also always be added for the victim. If parties other than the polluter are able to influence the accident risk, they might be held liable for the amount of their contribution to the damage. That, however, is not an argument against imposing strict liability on the polluter.

Applying the above insights for determining the choice between negligence and strict liability in environmental cases, there are strong arguments in favour of strict liability. Many cases of environmental pollution will be truly unilateral in the sense that it is only the injuring party's activity that has influenced the risk of accident, which points to strict liability. Nevertheless, in some cases it will be the victim's activity that caused the harm – for example, where the victim knowingly came to the nuisance. In the context of global warming and associated rising sea levels, victims should bear part of the damage if they move to areas prone to flooding: this may lead to the rejection of a claim for compensation.[19]

---

17 Adams (1989); Diamond (1974); Shavell (1980).

18 See Faure (1995).

19 See Wittman (1980).

A further important difference between negligence and strict liability should be mentioned. Negligence requires judges to set standards of care. The information required is typically costly and difficult to obtain. This makes negligence a costly basis of liability as courts have to expend resources to obtain that information; if they reach a decision based upon insufficient information, those decisions could be systematically incorrect. Strict liability shifts all costs to the injuring party, who will then define the optimal care level. If information on optimal precaution is more freely available to industry players, strict liability may be preferred.

If risk aversion is introduced and the potential injurer is risk averse, Endres and Schwarze show that strict liability is efficient only if risk can be removed from the risk-averse injurer, for instance, through insurance.[20] Furthermore, it is assumed here that judges have accurate information as to the extent of the damage. If courts err in assessing the level of damages, strict liability can lead to under-deterrence. If, by contrast, courts more easily observe the socially desirable level of precaution rather than the exact amount of external harm, a negligence rule is preferred.[21]

3.1 Environmental Liability versus Regulation

Although environmental liability is an important instrument for preventing environmental damage, it still finds little application in the context of climate change. Liability actions can be used to claim compensation for damages, to seek injunctive and/or declaratory relief and in procedural injury cases (such as challenges to environmental impact assessments and judicial review). Most claims are unsuccessful, and are directed against public authorities rather than industry. Successful claims tend to relate to procedural injury,[22] although injunctive or declaratory relief cases are potentially successful.[23] Climate change litigation is currently more of an issue in North America and Australia than it is in Europe.

Environmental liability has serious limitations. Some scholars therefore qualify the public law approach as 'the preferred approach' for preventing environmental damage.[24] If the conditions for the application of liability law are not fulfilled – such as when the victims and injuring party, or causal relationships between an activity and the damage, cannot

---

20 Endres and Schwarze (1991).
21 Cooter (1984).
22 Kosolapova (2011).
23 Spier (2011); Van Dijk (2011).
24 Bergkamp (2001).

be determined – instruments other than liability law will have to be used. In the context of climate change, the effective application of liability rules is impeded by a series of issues which will be addressed after concisely presenting the general law and economics insights.

Law and economics literature has closely examined whether safety in society can best be achieved through liability rules or through regulation. An important contribution comes from Shavell, who has examined the choice between the two options.[25] Both liability rules and regulation can be used to optimally prevent environmental harm, but their approaches differ. Tort law offers incentives for taking optimal risk prevention measures by inducing those who create risk to pay damages or compensation, but it leaves the choice of measures to the risk creator. Through regulation, the government imposes a standard *ex ante*. Non-compliance with this standard may lead to administrative or criminal sanctions. Regulation is thus considered an *ex ante* system aimed at prevention, whereas liability rules mainly intervene *ex post* after the damage has occurred. The criteria advanced by Shavell that influence the choice between safety regulation and liability are described below.

Information asymmetries

Deficiencies in information, as a cause of market failure and justification for government intervention,[26] are essential elements to be considered for the proper operation of a liability system. The parties in an accident setting generally have better information relating to the accident risk than do regulatory bodies[27] and are consequently better able to determine optimal preventive action. This assumption is reversed if the risks are not appreciated by the parties. This is especially a problem if the costs are external and cannot be assessed by the parties.

Insolvency

If the potential damages exceed the individual injurer's ability to pay, liability rules will not provide optimal incentives since the injuring party will consider the accident as having a magnitude equal only to his financial resources, and will thus take less than optimal precautionary measures.[28] This gives rise to under-deterrence.[29] Insolvency causes

---

25 Shavell (1984a), (1984b) and (1987).

26 Stigler (1961); Schwartz and Wilde (1979); Mackaay (1982).

27 Shavell (1984a).

28 Ibid.

29 Shavell (1986).

problems especially with rules of strict liability, but less so with negligence.

Environmentally oriented safety regulation can overcome this insolvency problem (as could insurance, provided that the moral hazard issue associated with insurance is reduced). Under safety regulation, the appropriate standard of care is determined *ex ante* and will be affected by enforcement instruments to compel compliance with the standard, irrespective of the injuring party's financial resources. If safety regulation is introduced to address potential problems of inability to pay compensation, non-monetary sanctions should be employed to avoid a recurrence of the problem.[30]

Under-deterrence of tort law

Some activities cause considerable damage, but are unlikely ever to end up in court. In the absence of litigation, liability rules will not be effective and regulation that sets efficient standards of care is preferable.[31]

An injuring party can escape liability where the damage to each individual is small (for instance, damage caused to common property), where the damage takes time to materialize and evidence has been lost, or where the injurer has gone out of business.

Proving causal links between the offending activities and the damage can also be difficult.[32] The burden of proving a causal relationship becomes more difficult as time passes; victims will often not recognize that the harm originated from a tortious act, and may think it had a natural cause. For such reasons, a legal proceedings may not be brought, making environmentally oriented safety regulation necessary to ensure that the potential polluter exercises sufficient care.[33]

Also in the context of climate change liability, Shavell's criteria constitute important points of departure. The issue of insolvency, for example, would become particularly important if courts were to accept joint and several liability in climate change cases. This could severely endanger the financial position of commercial entities.[34]

The effective application of liability rules can be impeded by high costs associated with legal fees, costs of obtaining information and evidence, and the risk of an erroneous ruling. Such costs limit the

---

30 Shavell (1985).

31 Shavell (1984a).

32 Landes and Posner (1984); Kunreuther and Freeman (2001).

33 Bocken (1987) and (1988).

34 Kaminskaitė-Salters (2011).

number of court cases being brought and give rise to under-deterrence. The requirement to prove a concrete and particularized injury as a precondition for legal standing will also negatively affect the prospects of an effective liability regime for climate change. Class actions or the possibility for non-governmental organizations (NGOs) to have standing may increase the number of liability claims. Even with class actions and legal standing for NGOs, the effectiveness of climate change liability is still undermined if liability cases – as is arguably to be expected – remain restricted to a limited number of victims and defendants. Liability rules, therefore, might not provide a structural approach to climate change.

Moreover, a liability approach may not have a sufficiently strong preventive focus for the pressing problem of irreversible climate change since, in many jurisdictions, claims (other than those for injunctive relief) may only be brought once the damage has actually occurred. Furthermore, liability claims might be weakened by the statutory authorization defence that a company that is complying with legal rules does not break the law and, therefore, should not be found liable. Another problem for a strong climate change liability regime may be the so-called political question doctrine whereby courts are reluctant to extend the law to areas that are perceived by judges to be too 'political'. As a consequence, courts may not welcome climate change liability claims. For such an example see Chapter 8.

Liability rules alone, therefore, are not sufficient to prevent environmental harm, particularly in the context of climate change. Instruments other than the command and control type of regulation, such as taxes or emissions trading systems, must be considered.

According to Shavell's criteria and the discussion above on climate change and liability, there is thus a strong argument for controlling environmental risk through *ex ante* regulation (or other appropriate instruments). However, the effectiveness of regulation depends upon enforcement, which may be weak and is susceptible to lobbying. Regulation also lacks flexibility and can become outdated. Based upon these considerations, liability rules are an important complement to regulation, for example, in relation to individual damage cases.[35]

One element that still needs to be considered with regard to regulation by liability rules is its acceptability. The acceptability of courts resolving conflicts may be very high in some jurisdictions, but not in others. For

---

[35] On the complementary relationship between tort law and regulation see Rose-Ackerman (1992) and (1996); Faure and Ruegg (1994); Kolstad, Ulen and Johnson (1990); Arcuri (2001) and Burrows (1999). For a comparison between the US and Europe see Rose-Ackerman (1995) and (1996).

instance, it is often reported that in the United States litigation is widespread and accepted; in Japan, by contrast, the opposite appears to be true. In societies where harmony is highly valued, liability rules may not be as acceptable and hence not work as an effective deterrence mechanism for polluters.

4. ENVIRONMENTAL TAXATION

Economic theory predicts that if an item is underpriced, more of it will be used. This proposition becomes particularly important if its 'excessive' use reduces the standard of living of market participants. Greenhouse gases did not have a market price and entrepreneurs did not take into account the negative effects they inflicted upon the environment (or 'negative externalities' as economists say). This 'market failure' can be overcome by bringing the item into the market price mechanism (that is, by internalizing the negative effects). The basic intuition behind this is that the price of the item should be inflated to adequately reflect social costs in order to create incentives to use less of the item. A tax scheme or an emissions trading scheme can be used to increase costs for producers. Whether producers or consumers will bear its burden will not depend on which of the two are legally obliged to pay the tax, but rather on which of the two societal groups can least avoid it.[36]

With regard to the introduction of a CO_2 price, a reduction in the growth rate of the gross domestic product (GDP), or even a GDP contraction, is to be expected. Despite this effect or rather, precisely because of it, society is better off. The absolute loss to society that stems from the excessive use of scarce resources is reduced to a socially desirable level. Since the GDP considers only economically quantifiable data, it does not take into account the destruction of the environment and is thus not a viable instrument for measuring social wealth.

The Pigouvian tax for environmental pollution was first developed by Arthur C. Pigou. This concept states that pollution can be reduced to socially desirable levels by increasing an entity's marginal private costs by means of a tax to reflect the marginal costs incurred by society (that is, costs reflecting all negative externalities). Since the production costs of polluters are increased by the tax to reflect the costs to society, firms maximize profits by reducing activity levels to the social optimum.[37] Some authors have recast Pigou's ideas in terms of the costs of pollution

---

36 As economists say, it depends on the income elasticity of demand.
37 Groosman (1999). See also Turner, Pearce and Bateman (1994).

abatement. Here, the marginal costs of pollution abatement for the firm are compared with the marginal external costs of abatement.[38]

Taxation has a number of advantages:

1. A carbon tax provides a clear and continuous incentive for abatement activities to reduce the tax burden. Taxing carbon emissions with a fixed price per unit provides an unambiguous price signal. Such a clear and fixed price signal mitigates business uncertainty concerning investment in innovation and in abatement technology. The absence of price volatility reduces the risks associated with abatement investment and this will also stimulate innovation in abatement technology.
2. Taxation generates income for the government. Since environmental taxes are used to internalize a market failure, they can be expected to give rise to less distortion of relative market prices than taxes imposed on labour or capital, for example.[39] These distortionary effects of taxation can, of course, be reduced.
3. Taxes can be adjusted if necessary to sharpen the environmental goal or alternatively to limit the cost burden on the industry. Such adjustments, however, may change the risk perception of firms and thus negatively affect their investment decisions in abatement technology.
4. Environmental taxes do not distort competition within the same jurisdiction in the sense that they do not give a comparative advantage to incumbent undertakings. The reason for this is that all undertakings – both existing and new entrants – are subject to the same legal framework. This positive aspect of taxation is an advantage that may arise with respect to those emissions trading systems that use free allocation based on historic emissions to allocate emission allowances, such as grandfathering.[40]

Despite the benefits presented above, environmental taxes are not free of problems, and give rise to a number of critical issues:

---

[38] See Pindyck and Rubenfield (2001). See also Perman et al. (2003), pp. 17ff.

[39] A tax on labour makes workers more expensive in comparison to machines. As a consequence, businesses hire fewer workers and buy more machines for production. Workers become less well off, while machine producers are better off.

[40] Grandfathering describes free allocation of emission allowances on the basis of historic emissions.

1. While the price burden of the carbon tax is known to undertakings, the environmental effectiveness of the tax scheme is uncertain. Emission levels among companies may vary, which may imply that environmental targets will not be reached.
2. A tax rate should be set in such a way that the marginal benefits of pollution equal the marginal costs of pollution. Failure to set an appropriate rate undermines the effectiveness of the tax in achieving a full internalization of the externalities. The Pigouvian tax is a simple incentive-based mechanism that is capable of inducing behavioural changes. To set an optimal tax rate, however, governments require detailed information, which will not always be available.[41] If the tax is set too high, companies may downsize or even shift production abroad. Similar concerns arise with emissions trading systems that employ too stringent targets. Here, the shifting of production abroad is referred to as carbon leakage.
3. The effectiveness of the tax also depends heavily on demand and supply. Only when the demand and supply curves are flat (that is, elastic) will a price increase lead to a significant change in consumption or supply patterns and make the tax effective. If either supply or demand, or both, are highly inelastic, the effect of a tax will be very minimal; it will then not change consumption or production patterns.
4. The use of a flat rate tax is not fully consistent with the idea of the Pigouvian tax. As companies have different marginal cost functions, a flat rate will not provide appropriate incentives for emission reductions for all.
5. In general, taxes have regressive effects on some income categories, and a carbon tax is no different. Although, from an economic point of view, it is highly debatable whether such income effects should be included in the efficiency analysis, from a political perspective such issues must be considered appropriately.

5. EMISSIONS TRADING

The absence of flexibility and incentives is often considered to be a disadvantage of traditional command and control approaches, which are too general to allow for a differentiation between the abatement potentials of polluting firms. Worse, fine-tuned regulation, such as traditional

---

[41] See Fullerton, Leicester and Smith (2010) for an elaborate criticism.

permit schemes, require expensive and burdensome bureaucratic work. This could be overcome by the use of increasingly popular marketable permits, which were introduced by Dales.[42] Under an emissions trading system, marketable emission allowances are issued to polluters up to the amount predetermined by the policy makers. The market price and the willingness of the polluter to pay determines which actors will pollute and which will invest in abatement technology. Many different distribution schemes are possible. Allowances can, for example, be auctioned or distributed for free.[43] In light of the previous discussion regarding environmental taxation, this section addresses emissions trading by, first, examining its advantages, and subsequently reviewing its critical issues.

5.1 Advantages of Emissions Trading

Marketable allowances have a number of advantages.

1. Emissions trading allows the regulator to determine the quantity of emissions in the economy by limiting the availability of emission allowances. The environmental effectiveness is thus directly determined by the legislator.
2. Emissions trading allows operators with different abatement costs to trade and thereby enables entrepreneurs to decide for themselves who will actually abate. This enables the scheme to reduce emissions at the lowest cost possible. According to the European Commission's own assessment,[44] the cost to the EU of its climate policy under the Emissions Trading System (the EU ETS) will be between €2.9 to €3.7 billion. In the absence of this trading system, the environmental costs would amount to €6.8 billion. The costs that are saved by using an emissions trading system are thus substantial.
3. Since emissions trading allows for the market price to be determined by demand and supply, it automatically adjusts for inflation.[45]

---

[42] Dales (1968).

[43] For a comparison, see Weishaar (2007a) pp. 36ff.

[44] European Commission (2005), p. 8.

[45] The problem of inflation is also recognized in an environmental taxation context. Many environmental taxes are adjusted for inflation.

4. Marketable permits may be more acceptable for regulators since they are already familiar with permits (with which emission allowances may be compared). Taxation, by contrast, is generally disliked by citizens.
5. The price in the emissions trading system is determined by demand and supply. If demand increases as a result of a growing economy, the price of emission allowances will also increase. This helps in preventing the economy from becoming overheated. In times of low or negative economic growth, demand declines and the price of emission allowances falls. When the economy is in distress the financial burden is alleviated. Emissions trading thus functions as an automatic stabilizer that dampens market fluctuations.
6. Cao argues that a cap-and-trade system is preferable in light of the common but differentiated responsibilities of developing countries. The reason for this is that permit trading is more flexible than a carbon tax since it allows for a more equitable allocation of emission allowances across states.[46] In this way countries with limited historic responsibility of climate change or with limited economic means to abate emissions can be enabled to continue to emit more.

5.2 Critical Issues

Marketable permits also have a number of disadvantages, which are listed below.

1. The volatility of allowance prices creates business uncertainty over future price developments. This can result in sub-optimal investment in innovation (research and development) and investment in abatement technology, or to limited investment in sustainable energy supply, particularly renewable energy. An emissions trading system may therefore produce disappointing results if charged with other policy objectives, such as incentivizing innovation, unless, of course, the system is expressly designed to address these policy objectives.
2. Setting the optimal emission level for an emissions trading system requires detailed information that is unavailable in the context of scientific uncertainty. Policy makers are thus required to take a pragmatic decision in setting emissions targets. Moreover, the

---

[46] Cao (2011) p. 22.

opinions of states differ over how much they are responsible for climate change and how large their contribution towards mitigation should be. There is, however, an emerging agreement between states on the overall limitation on temperature increases. More than 140 countries have endorsed the IPCC's Fourth Assessment Report with a view to reducing global emissions in order to hold the global temperature increase at 2°C.[47] However, countries still differ in their assessment of the effectiveness and the stringency of the required policy instruments in order to achieve that goal. These differences translate into very different national climate change targets that are reflected in the various emissions trading designs and targets. This can be seen in Chapter 4 where several systems are described.

3. Permit trading is associated with costs for polluters. As is the case with other instruments, polluters could try to avoid these costs by transferring production (and hence pollution) to another region with less stringent emissions reduction goals, or none at all. This is referred to as 'carbon leakage' in the context of emissions trading.
4. With emissions trading, polluters' profits may increase while pollution levels remain the same. This situation occurs when polluters receive emission allowances for free (such as under a grandfathering allocation format) and, as a result of inelastic demand, they are able to pass on a significant part of the carbon costs to consumers. This is commonly referred to as 'windfall profits'. Even though this only constitutes a 'transfer' from an economic perspective, windfall profits are politically undesirable since they are difficult to accept for the general public.[48] Windfall profits are discussed in Chapter 5.
5. Emissions trading systems that use free allocation do not raise revenue for the government, although revenue is raised if auctioning is used as an allocation mechanism.
6. The market price of emission allowances may be inelastic. Relatively small changes in the quantity can lead to a large price change. The market price of allowances, therefore, may be volatile. Price volatility is a normal feature in various markets and is not a point of concern unless the market price is intended to incentivize behavioural change or an economic transition. In such situations price volatility may add to business uncertainty and reduce the level of desired investment.

---

47 Copenhagen Accord (2009) para. 9.

48 Woerdman, Couwenberg and Nentjes (2009).

7. Emissions trading is associated with transaction and monitoring costs. Benefits of cost savings will be sufficiently large to outweigh the transaction costs for large emitters. For small emitters, emissions trading may thus not be feasible. An effective way to reduce monitoring and administrative costs is by reducing the scope of the trading system. Next, the scheme may focus on process emissions and on fuel inputs in production processes. Measuring emissions from both, however, is costly, although in differing degrees. Data on fuel inputs and energy use is more easily obtainable and allows the internalization of at least some of the environmental costs. Such an approach – albeit sub-optimal in the context of large processing plants – could still be desirable when process emissions are limited because of the absence of large plants (in city trading schemes, for example).
8. If small emitters are subject to other successful policies that target emissions (such as, for example, energy efficiency standards or building standards), this can have repercussions on emissions within the emissions trading system. Less energy consumed by households may suggest that fewer emissions are generated by installations covered by the ETS. If this is the case, successful emission reductions achieved by those associated policy fields will permit more emissions within the ETS than were probably expected *ex ante* (the 'waterbed effect'). This effect may also arise, for example, if the energy sector is covered by an emissions trading system while effective policies to support renewable energy are also in place.

6. CONCLUDING REMARKS

The above treatment has shown that there is no single instrument that is optimal to internalize environmental externalities in the most effective way. Depending upon particular circumstances, the instruments discussed in this chapter all have their specific strengths and weaknesses: no system is perfect. They will therefore normally be applied in conjunction with each other. While climate change liability is still a remote possibility in most jurisdictions, taxation and emissions trading are in practice frequently found to be employed in conjunction with each other. A few words from an ETS design perspective on the relationship between the two schemes are therefore in order.

Flat carbon taxes are – depending on the monitoring system – easily applied and have a limited administrative burden. They are, however, not capable of differentiating between the abatement costs of polluters and thus give rise to higher overall carbon abatement costs. They may therefore be sub-optimal for large emitters where the foregone benefits of additional reductions in abatement costs outweigh the benefits of lower administrative costs. This may be especially relevant for large installations.

In contrast to the above, emissions trading systems allow polluters to trade and thereby to abate at the lowest possible costs. Administrative costs for polluters may, however, be higher than those under a flat tax scheme since polluters must monitor market prices and decide if and when to buy or sell allowances. That may, therefore, be less than optimal in those situations where the disadvantages stemming from higher administrative costs outweigh the benefits of additional reductions in abatement costs. This may be especially relevant for small installations.

It therefore appears to be sensible to run both a (carbon) tax and an emissions trading system in parallel. The tax system could focus on small installations, which would give rise to only a limited degree of reduction in abatement costs, while an ETS could be used for large installations where the cost advantages from permit trading between polluters are large. Having a greenhouse gas tax scheme apply to operators that are also subject to a greenhouse gas emissions trading scheme will increase the administrative burden and negatively affect the cost effectiveness of abatement. If policy makers are able to achieve their objectives by introducing only one instrument rather than two, this is the preferred cause of action.[49]

Administrative costs are an important concern when designing a policy instrument and when deciding between opting for a tax scheme or an emissions trading system. It is therefore important to point out that the monitoring and enforcement costs of an emissions trading system can be lowered by focusing only on fuel inputs that can be easily obtained, for example, in the form of energy bills. This, of course, only allows the internalization of the externalities of greenhouse gases that stem directly from those inputs. Emissions associated with production processes are not accounted for. This clearly constitutes a second best solution, but it may prove efficient where the ETS focuses on sectors or areas that do not have many process emissions.

A further element is that emissions trading is often only one instrument in a wider policy of climate change and energy. Optimal mixes between the applied instruments in the related policy fields are essential. A

---

[49] Weishaar and Tiche (2013).

successful energy policy substituting coal-fired power plants with wind energy will, for instance, reduce covered emissions by power plants. With the reduction of covered emissions, the price of emission allowances will also fall. This, in turn, may enable other covered sectors to emit more (the waterbed effect). If there is a strong interaction effect between various policies, it is wise to take those into account when designing an emissions trading system; these interactions will be discussed further in Chapter 3.

3. Design variants of emissions trading

1. INTRODUCTION

Emissions trading systems (ETSs) and their design variants have been subject to considerable research. The literature in this area is rich with studies comparing the various types of system. This observation seems to corroborate the fact that academics are very much intrigued by the special situations in which one emissions trading system comes out on top of another system or vice versa – in other words, under which circumstances one system is to be preferred over another.

Academic typologies and findings, however, appear to have limited practical impact on emissions trading designers. It appears that the research interests of academics are sometimes too far removed from the policy considerations and trade-offs faced by policy makers. This chapter, therefore, does not follow the conventional academic approach by starting from the literature and describing the generic emissions trading models that exist. It rather follows the problems encountered by ETS designers: after all, the problems that policy makers are required to deal with are even more complex than the design choices normally considered by academics.

The natural drawback of this approach is that it cannot provide clear guidance to policy makers of what they should or should not do. The reason is straightforward: policy advice must be given for the particular context in which the system will operate and given the nature of the various policy objectives. This chapter provides food for thought with regard to specific trade-offs that ETS designers will face if they seek to achieve several policy objectives simultaneously – something which policy makers typically tend to do. Emissions trading designers will thus see that in some situations a particular design could be promising, while for other objectives another model may be more beneficial. This revelation should then lead designers to tap into the academic literature with clear questions in mind. Only with such clear questions will they be able to find the answer they are looking for since there is no 'one size fits all' ETS design.

To paraphrase the above, we will draw upon Gary T. Marx[1] who (in a very different context) was able to convey similar ideas in a more eloquent way:

> While some philosophers may lust after a Rosetta stone of clear and consistent justifications, a central argument here is that in matters so complex and varied we are better served by an imperfect compass than a detailed map. Such a map can lead to the erroneous conclusion that ethical directions can be easily reached or to a statement so far in the stratosphere that only angels can see and apply it. Maps of uncharted terrain are hard to come by. A chart of new territories needs to begin with simple coordinates and rough estimates.

This chapter is structured as follows. It first describes, in Section 2, the underlying policy framework: the context within which an emissions trading system must operate. It is, after all, the particular circumstances that determine which design features are likely to work or not, and therefore which instrument should be chosen. Section 3 then examines the policy objectives that emissions trading designers may wish to achieve and the possible design choices in light of efficiency, environmental effectiveness and acceptability. Section 4 presents the most common design choices that are available to ETS designers: the 'nuts and the bolts' that designers can employ to forge their own system. A conclusion highlights our main points.

2. POLICY CONTEXT

Any emissions trading system will need to operate within a given context. This context is co-determined by a large number of political, social and economic factors. The section below presents a series of issues that ETS designers should consider when thinking about their system designs and when setting their policy objectives. The underlying assumption here is that the general policy framework will impact on how an emissions trading system will eventually function and whether the set policy objectives can be achieved with a particular design. We have chosen to present the identified issues along the three meta-dimensions of political factors, economic factors and social factors. The list of issues is, of course, not exhaustive, but we present the most important of them and will provide designers with a good idea of the relevant issues.

---

[1] Marx (1998) p. 182 (the citation has been adapted slightly).

2.1 Political Factors

The Intergovernmental Panel on Climate Change (IPCC) has offered guidance on the greenhouse gas (GHG) emission reductions that are required to prevent dangerous climate change. Scientific uncertainty over those targets necessarily remains because scientists work with models that are only as good as the assumptions upon which they are based. Scientists work with confidence intervals and are able to offer advice only on the basis of a (very) high level of certainty – nevertheless, some uncertainty always remains. It would appear to be prudent, therefore, to act bearing in mind the precautionary principle and act decisively. When setting (stringent) targets to prevent global climate change, policy makers should bear in mind that they will also need to take into account the global collective action problem in that some countries will not honour their responsibility to reduce emissions. In such a context, target setting is inherently difficult. Since reducing GHG emissions will generally lead to a lowering of economic prosperity, as measured in terms of GDP,[2] the targets set will be subject to domestic debate. To produce the desired investments by entrepreneurs, policy targets and political commitments need to be credible.

Policies, therefore, must be well designed and thought through. Emissions trading as a policy instrument requires a significant amount of work. Even though the textbook model of an ETS appears to be easy to implement, it needs a significant amount of careful planning, with the collection of data and enactment of legal rules. All of this requires time. If time is not available to gather sufficient accurate data on the actual emissions of installations that are to be covered, a fully fledged emissions trading system is probably not a good policy choice. The reason is straightforward. Greenhouse gas emissions may be based on fuel inputs but they also derive from production processes. Around 55 per cent of the GHG emissions of a cement production plant are derived from the production processes;[3] emissions from these processes may be difficult to monitor and measure. Fuel inputs, by contrast, can be obtained easily by examining energy bills of fuel purchases and multiplying the consumed quantity of a fuel input with its estimated global warming impact

---

2 Gross domestic product is generally criticized as an imperfect indicator for economic output since it does not consider, e.g., charity work, the work performed by homemakers, or environmental degradation. It is therefore a misleading indicator when it is used as a proxy for the overall well-being of a country.

3 Williams (2007).

(determined in terms of tonnes of CO_2 equivalents). If there is insufficient time to gather accurate process emissions data, it may thus be expedient to cover only the costs of fuel input emissions until the data is available. Even though this will give rise to a sub-optimal emissions trading system, it may provide for more political acceptability. In situations where the ETS barely covers installations with many process emissions (such as in metropolitan areas with relatively less heavy industry), this may be a feasible option.

Policy makers are charged with the task of weighing various societal interests against each other and to take decisions. In the area of climate change this is also the case and frequently there will be various stakeholder interests at play. Of particular importance, however, are also the various policy areas that will affect climate policy. When designing an emissions trading system, issues such as energy, construction standards and industrial policy will also be implicated. Policies in very diverse areas may interact with, or depend upon the working of, the emissions trading system. The specific design chosen for the system may lend additional support for those other policy areas, or could actually undermine them.

An example illustrates this point. The European Union (EU) has set '20-20-20 goals'[4] for smart, sustainable and inclusive growth with which the EU addresses all areas needed to mitigate global warming. It leads to a reduction in allowances per person (energy efficiency), it addresses the emissions intensity of the economy (via the EU emissions trading system), and it greens energy production (renewables).

While addressing all areas that contribute to climate change makes intuitive sense, it must be observed that the three policy areas falling within the 20-20-20 goals affect each other. More renewables means that less energy production takes place within the framework of the EU ETS, which reduces the market price for emission allowances. Similarly, if private homes need less heating, energy demand is lower and there will be less demand for emission allowances by energy producers covered under the EU ETS. Since the EU ETS is designed as a cap-and-trade system, it also means that other covered sectors will face lower allowance prices and thus have fewer incentives to invest in abatement technology. Reductions in renewables and energy efficiency are thus partially outweighed by additional emissions in the industry sector. This interaction

---

[4] The 20-20-20 goals set three key objectives for the year 2020: (i) a 20% reduction in EU greenhouse gas emissions from 1990 levels; (ii) increasing the share of EU energy consumption produced from renewable resources to 20%; (iii) a 20% improvement in the EU's energy efficiency.

effect, however, is not a one-way street. It works both ways since higher emission allowance prices incentivize investment in energy efficiency and renewables.

In light of the low price of emission allowances in the EU ETS, it may not be surprising that the European Union is not on track to meet its renewable energy and energy efficiency goals. The 20-20-20 goals for energy efficiency have failed by a large margin: the abatement that should come from energy efficiency improvements was estimated at 368 million tonnes of CO_2 equivalents, while it is projected that only 206.9 million tonnes will be achieved.[5] The renewable energy targets may still be attainable if the private sector is prepared to invest an additional €70 billion until 2020.[6] Recently, policy makers seem to have put a different emphasis on the underlying policy objectives of the EU ETS and have started to criticize it for not delivering policy objectives that it was not designed to address. This is unfair since policy instruments tend to perform best in those tasks they are designed to address. Emissions trading designers should therefore be very clear about the wider scope of their policy objectives.

To emphasize the importance of this issue, it should also be realized that it is not enough to determine the wider policy objectives, assess the interaction effects between them and design an emissions trading scheme that functions well. Orchestrating a complex policy field means also that the policy mix needs to stay in tune in the future. It is thus not simply a static concept; it also has a dynamic aspect. Introducing new rules on 'fracking' to permit the large-scale use of shale gas,[7] for instance, will have significant consequences for the energy mix in a country, and therefore in its greenhouse gas emissions (shale gas is cheaper than coal and contains only half as much GHG emissions per unit of energy). This means that the large-scale application of this new source of energy will greatly reduce the price of emission allowances. Unless the emissions trading system is designed to adapt to the oversupply of allowances that

---

5 European Commission (2012d) p. 3.

6 European Commission (2012a).

7 Shale gas is obtained from shale formations. Since this gas is not contained in large reservoirs, the shale needs to be fractured in order to obtain it. The fracturing is done by pumping Fracfluid (consisting of water, sand and chemicals) under high pressure into the shale. The exploitation of shale gas is economically attractive and is rapidly increasing in the US. There are, however, strong environmental fears that fracking could, among other things, pollute the ground water.

is likely to arise, the system may become ineffective in incentivizing investments in green technology.

A further issue is that, although the legislator may put a price on emissions, firms and consumers may try to avoid the costs and thus impair the effectiveness of the scheme. Firms can 'vote with their feet' and leave, and consumers can purchase products that were produced abroad without taking account of GHG emissions. These phenomena are commonly referred to as 'carbon leakage'. Although it is a clear theoretical possibility, carbon leakage is projected to occur only at high allowance prices and to a more limited extent than entrepreneurs would commonly assert.[8] Since enterprises offer employment to voters, they constitute an important force that can have a bearing on the shaping of policy objectives, which may then translate into ETS design issues.

2.2 Economic Factors

ETS design must also take into account the overall industrial structure. With regard to entities that emit many process emissions or only emissions derived from fuel inputs, it was discussed above that a simpler system that addresses only fuel emissions may be more efficient in specific circumstances. Designing complex laws to capture a few emissions from very few installations may not be economically justifiable: administrative costs may exceed the gains in terms of lower abatement costs.

Industry structure is important in other respects. If the emitting industry is concentrated mainly in sectors that do not add decisively to economic output and employment, and therefore carry less weight in terms of lobbying and the political decision-making process, it is possible to move forward quickly. In the North American Regional Greenhouse Gas Initiative (RGGI) – a sub-national emissions trading scheme in the United States that regulates electricity producers – auctioning of emission allowances was introduced to curb greenhouse gases. Since the RGGI states do not have a sizable heavy industry, electricity producers were subjugated to an emissions trading system which applied auctioning instead of grandfathering as an allocation rule. The trading system thus provides income that can be distributed back to voters. Such a 'deep

---

[8] Mattoo, Subramanian, van der Mensbrugghe and He (2009); Heilmayr and Bradbury (2011). Although it must be observed that with emissions trading systems that simply regulate electricity, carbon leakage may occur also at lower CO_2 price levels as the transport costs of energy are relatively low.

pockets approach' (an ability to pay approach) can be an important consideration for policy makers when designing an emissions trading system.

Related to the market structure is also the question whether or not situations of market domination or cartelization are likely to arise. The industrial economic concept of market power that requires dominance or joint dominance, or a highly concentrated industry to facilitate cartelization, is probably unlikely to arise if the coverage of an emissions trading system is large. Market power may, however, arise also in the context of emission allowance auctions or on the secondary market in the form of market manipulation or market cornering. Dominance in the industrial economic sense is not required for problems to arise. If a particular company accounts for vast amounts of emissions, it may be able to influence the allowance market.

An important and frequently disregarded feature of emissions trading systems is that they give rise to inelastic markets. A small decline in output then triggers strong price reductions in emission allowances, while at times of economic prosperity the price rises sharply. Market prices may thus be quite volatile – at times very low, while at other times they may be very high. Volatile market prices have been criticized for sending an unclear signal to investors, undermining the trigger for meaningful investment in abatement and innovation. Such volatility is especially problematic with regard to investment in costly infrastructural projects that reach their break-even points after many years. These will be very risky business decisions, indeed. Price volatility does not affect the (static) environmental effectiveness of the emissions trading scheme, but may – depending on the policy objectives – still require support schemes for limiting carbon leakage or containing price volatility.

Emissions trading schemes are market-based instruments. They allow enterprises with the lowest abatement costs to abate more than firms with higher abatement costs. This characteristic constitutes the very basis for achieving cost-effective emission reductions and sets emissions trading apart from other instruments in the policy mix. The anticipated benefits from emissions trading may, however, not materialize if enterprises do not take their abatement decisions on the basis of their abatement costs, which may occur if enterprises are influenced by policy considerations. In a Japanese context, the term 'Iron Triangle' describes the amicable relationship between industry, bureaucracy and politicians that lead entities to take decisions that are in their best long-term interest, albeit not always rational from a purely commercial perspective. Clearly, nowadays, the Iron Triangle is but a shadow of its past, and is possibly no longer very influential in Japan's contemporary business culture. But there are other examples. In China many enterprises that are covered by the six emissions trading pilot schemes are state owned enterprises (SOEs). State owned enterprises can be influenced by local government and may not be entirely free to take their abatement decisions on the basis of their abatement costs alone. This means that an SOE may need to abate or purchase allowances from another company on terms that an independent entrepreneur would not have been willing to accept. As a result, emissions abatement is not undertaken by the installations with the lowest abatement costs, and/or inefficiencies stem from subsidizing inefficient companies. The efficiency gains associated with emissions trading will not materialize, or will materialize to a lesser extent. In the presence of such circumstances, the application of an emissions trading system must be carefully analyzed and its expression found in the particular design features of a system that is capable of restricting administrative guidance.

Another impediment to the free working of market forces that ETS designers should bear in mind is price regulation. If enterprises covered by a scheme are barred by law from increasing their prices, emissions trading will be unable to trigger behavioural changes in consumers. This, however, is desirable since it is ultimately thc consumer who demands too much of a product that is underpriced. In a context in which energy prices are regulated – as is the case in China[9] – it may be difficult for state owned enterprises to pass on carbon costs to consumers. Passing on carbon costs would, however, induce consumers to purchase more energy-efficient electrical devices, invest in insulation and so on, and thereby reduce their overall carbon footprint.

Emissions trading is based on the exchange of allowances, and therefore needs efficient and cost-effective markets in which to trade allowances. These generally take the form of 'exchanges' where allowances are traded in real time. Where such exchanges do not exist in some jurisdictions, they would need to be established since mere 'over the counter (OTC) trade', where sellers and buyers must first find each other, may prove to be costly and thus undermine the efficiency of the system. In the context of the EU ETS, the early transactions took place in the form of OTC transactions and on exchanges. Derivatives emerged shortly thereafter. It is estimated that at present the vast majority of transactions in fact take place on the derivatives market. Financial products surely offer the possibility of hedging risks and it is reported that energy companies only revert to trading on the allowance spot market if

[9] See Jotzo (2013) pp. 35ff.

allowances are cheaper there. But – as the EU example shows – such financial products are not a precondition for the operation of an emissions trading market. This observation may be relevant for market environments where financial products are strongly regulated.

2.3 Social Factors

In addition to the political and economic factors that define the framework of operation of an emissions trading system, there are social factors that will determine its effectiveness. Issues discussed here relate to the trust in the rule of law, in regulatory certainty, as well as in monitoring and enforcement.

Emissions trading is a market-based instrument. To operate, it requires trust in the market. This trust must come from policy makers and enterprises alike. In addition, it requires a strong legal framework that not only allows enterprises to attain their rights vis-à-vis other enterprises, but also vis-à-vis the supervising authorities that govern the emissions trading system. The legal framework needs to afford entrepreneurs a sufficient degree of legal and regulatory certainty to induce them to invest in abatement technology and to take part in the system. If such a legal framework is insufficiently developed, an emissions trading system is unlikely to realize all the benefits that it could generate.

It is not only with regard to legal certainty and a stable investment climate that an ETS requires trust. Trust in the correct monitoring and enforcement of the system must also be present. If the execution of monitoring, reporting and verification – which is frequently based on self-monitoring and self-reporting – is widely regarded as being affected by criminal activity or corruption, it is very unlikely that an emissions trading system would be able to produce cost-efficient abatement. It might still lead to some abatement in greenhouse gases, but this might have been better achieved through the use of other instruments that may be less prone to corruption and criminal activity. Chapter 6 below addresses the value added tax (VAT) frauds and the 'phishing' attacks from which the EU ETS was suffering to such an extent that trade on the entire emissions market was suspended.

2.4 Conclusion

This section has reviewed several factors that impact on the context within which an emissions trading system must operate. Emissions trading designers are well advised to carefully consider the factors that affect the environmental effectiveness, efficiency and acceptability of a trading system before starting to make design choices. The political, economic and social factors discussed above are important examples of relevant issues to be examined, but they should foster the understanding that there is no 'one size fits all' ETS that is perfect for all situations and countries. Like any legal transplant – and emissions trading systems are strongly embedded in a legal regime – a 'copy and paste' is not possible. Any emissions trading scheme must be calibrated to the particular needs of the jurisdiction in which it is intended to operate. This is particularly true since the political objectives that are to be achieved by the emissions trading system may differ from jurisdiction to jurisdiction and are dependent on the larger policy framework in place. The next section considers potential policy objectives and which design features may be helpful in achieving them.

3. POLICY OBJECTIVES

After emphasizing the importance of closely examining the larger framework within which an emissions trading system will operate, our interest shifts to policy objectives. Many academics working on emissions trading will happily attest that greenhouse gas emissions trading is about cost-effective abatement. While this is certainly the most appealing property of such systems, it is often not the only policy objective with which a system is charged. The reason for this is that the world of the policy maker is far more complex than to resort simply to one policy objective; there is a need to balance many different societal interests.

The section below outlines different policy objectives and shows how different design solutions are assessed from the perspectives of environmental effectiveness, efficiency and political acceptability. The more policy objectives that ETS designers hope to attain, the more difficult it will be to find an appropriate design solution. As the reader will notice, at times the design solutions directly contradict each other. This may explain why emissions trading design is such a complex matter and why the academic literature does not seem to offer concrete solutions for designers.

The policy objectives reviewed here include environmental effectiveness, cost effectiveness, raising government income, economic growth, and incentivizing investment in clean technology. It goes without saying that this list is by no means exhaustive.

3.1 Environmental Effectiveness

Environmental effectiveness means that a predetermined greenhouse gas reduction target is achieved. Consequently, whether a target is ambitious or lax is not considered; it matters only that the objective has been realized. Obviously, setting the correct target is important, but that part of the discussion is omitted here.

The easiest way to ensure that an emissions target will be reached is to use a cap-and-trade system in which a predetermined amount of allowances is distributed to emitters. Since emitters are not allowed to emit more greenhouse gases than the amount allotted in emission allowances, the environmental effectiveness of the trading system is automatically safeguarded (provided, of course, there are sufficiently high penalties for defaulting, and effective monitoring and enforcement).

Emissions trading systems operating in the absence of a fixed emissions cap cannot automatically ensure environmental effectiveness. A credit-and-trade system could mean that industrial output could increase even beyond the desired emission levels. In the absence of a fixed and legally binding cap, these trading systems could still be environmentally effective if production benchmarks are tightened once the 'intended' emissions target is approached. In order to ensure the attainment of this target, policy makers must react ahead of time to tighten the standards and have tools at their disposition to induce sufficient and timely investment in emissions abatement. This requires a substantial degree of political determination, foresight and regulatory power that may not always be present. If these properties and capabilities are not available, environmental effectiveness cannot be safeguarded.

In any event, it can be observed that, in the absence of a legally binding cap, an 'intended emissions target' would need to include a 'security valve' when the setting of more stringent production standards is triggered to avoid emissions exceeding the intended target. The societal costs of maintaining an 'intended' emissions target that is reinforced by a 'security valve' involves higher abatement costs than a cap-and-trade target that is set equal to the 'intended' emissions target. This tightening of the 'intended' emissions target by means of the security valve makes such a credit-and-trade scheme more costly to society, but it could yield more emissions abatement. Since entities covered by the emissions trading system may be surprised by a 'sudden' sharpening of the emissions standards, their ability to plan investment in greenhouse gas abatement technology may be limited. This may place undue costs on them.

A more stringent emissions target does, however, allow emission-efficient enterprises to expand production and to gain market share as they will be able to minimize their abatement costs more than others. This, in turn, should have a positive effect upon competition and social welfare.[10]

Acceptability of the policy target depends on the particular design of the emissions trading system. Emissions trading caps will be acceptable to some stakeholders, but not to others. In general, one may assume that the absence of an emissions cap will be more easily accepted by industry, since it allows production increases, while an emissions cap renders the expansion of output more difficult. While an emissions cap was established in the EU ETS, the same was not possible in New Zealand. The New Zealand ETS has neither an emissions cap nor any quantitative restrictions on the import of Kyoto credits (although the latter is changing since New Zealand will have only limited access to Kyoto credits as it is not a member of the second Kyoto compliance period).

3.2 Cost Effectiveness (Lowest Abatement Costs)

The objective of 'cost effectiveness' of an emissions trading system seeks to ensure that greenhouse gas abatement occurs at the lowest possible cost. This involves a liquid market for emission allowances. Market liquidity is enhanced by enlarging the scope of the emissions trading system to more sectors of the economy or by linking the system with other emissions trading schemes. It also means that transaction costs for exchanging allowances are low so as to enable as many exchanges of allowances as possible. In addition to allowing access to low abatement opportunities in other sectors and abroad, and low transaction costs, administration costs should also be kept low. Such costs stem, inter alia, from monitoring, reporting and verification (MRV). These costs relate to the particular scope of the emissions trading system and the question of whether it is just fuel costs or other process emissions that should be covered. It has been suggested that credit-and-trade systems have high costs because they require each credit generation to be approved by a regulatory authority.[11] The economic effect of different transaction costs associated with different allocation mechanisms limits the degree of

---

[10] Weishaar (2009) Section V.2.2.

[11] Tietenberg, Grubb, Michaelowa, Swift and Zhang (1999) pp. 26, 34 and 107.

allocative efficiency[12] by reducing the range over which mutually beneficial exchanges are possible.[13]

3.3 Raising Government Income

At times of tight budgetary constraints, raising income by targeting 'public wrongs' can be appealing. Placing a price on greenhouse gas emissions means that enterprises will take the costs of pollution into account when making their production decisions. The income raised is correcting a market failure (namely that pollution was formerly without a cost). Thus, it does not distort the prices of goods and services. As policy makers do not know the abatement costs of covered entities, they should leave price setting of allowances to the market. This can be achieved via auctioning of allowances.

Emission allowance auctions allow the market participants themselves to determine the price of the allowances. Auctions can thus provide for efficient allocation. Moreover, Cramton and Kerr[14] find that auctions are appealing because auction revenues can be recycled to reduce distortionary taxes (the 'double dividend hypothesis'[15]). Investment will not leak abroad if the auction proceeds are redistributed back to businesses in the form of tax cuts (for example, by reducing distortionary labour taxes)[16] in such a way as to fully compensate individual operators for the expenses they incur.[17]

Auctioning allowances necessarily means that not all allowances can be allocated free of charge. Free allocation on the basis of benchmarking or grandfathering reduces the number of allowances that can be auctioned by the government and hence reduces government revenue. Covered entities also require a shortage of allowances in order to be induced to pay higher prices at auction. A cap-and-trade system is better able to

[12] Frank (1997) p. 350 defines allocative efficiency as a condition in which all possible gains from exchange are realized. This implies that those market participants valuing an item must have been able to attain it.

[13] See Harrison and Radov (2002) p. 52.

[14] Cramton and Kerr (1999).

[15] On the double dividend see, for example, De Mooij (1999).

[16] On labour taxes see Parry, Williams and Goulder (1999).

[17] The distortionary taxes on labour or profit could be substituted by the levy of non-distortionary taxes on CO_2 emissions. Ballard, Shoven and Whalley (1985) estimate that $1 raised through distortionary taxes can cost society $1.30. Thus society, including operators, could be made better off by auctioning. Such overcompensation through reducing distortionary taxes is also found by Bovenberg and Goulder (2000).

create this shortage than a credit-and-trade system and is therefore the preferred design choice for maximizing government revenue.[18]

Auctions may, however, not be the best choice in terms of public acceptability if vested interests are much in favour of free allocation. In particular, grandfathering is prominent, as it affords decision makers greater political control and allows distributional arguments to be taken into account.[19] Yet there are also examples where political support was in favour of auctioning. In the case of the EU ETS, for example, consumers suffered from high energy prices resulting from the passing on of opportunity costs of emission allowances to the end consumers. Because energy companies received their allowances for free, this led to 'windfall profits' for the electricity sector. Consumers were very displeased and auctioning was introduced. (Windfall profits are revisited in Chapter 5.) A further example, as mentioned above (in Section 2.2), is the RGGI system where auctioning was introduced to levy money for redistributive purposes.

3.4 Economic Growth

Emissions trading places a cost burden on undesirable activities. As a result, the formerly cost-free pollution of the environment is 'internalized' – that is, brought into the market price mechanism. As producers take the real costs of pollution into account, product prices increase and there is less production; hence the economy contracts. Reconciling emissions trading with economic growth is therefore a daunting task. Yet economic growth provides for prosperity and employment and is therefore an important policy objective that will be shared by many politicians.

There are various approaches to ease the economic burden of emissions trading on enterprises. These approaches include free allocation, price ceilings and trigger prices, offsets, support schemes and intensity targets. Each is considered in turn.

Free allocation

Within a cap-and-trade system emissions trading designers can choose to allocate emission allowances for free, for example in the form of

[18] Introducing auctions into a credit-and-trade system is theoretically possible but would defeat the object and purpose of the scheme: see Weishaar (2012).

[19] Stavins (1997).

grandfathering. Grandfathering allows incumbents to continue to produce, but they will generally seek to price the increased production costs (opportunity costs) through to consumers. This is rational business behaviour and their success is dependent on the demand price elasticity they face. Even though such allocation formats mitigate the direct out-of-pocket expenses for businesses, they do not get rid of the cost burden completely, and therefore limit economic growth. This is, after all, the desired result of the emissions trading system.

Price ceilings and trigger prices

Economic growth is facilitated by containing carbon prices and by allowing better access to emission allowances. One way of limiting carbon prices is by creating more emission allowances when the market price reaches a certain trigger price. This is commonly referred to as a price ceiling. A price ceiling system is employed in the Australian emissions trading system (discussed in Chapter 4). If the legislator does not wish to offer unlimited allowances at the ceiling price, it can also decide to sell limited quantities of allowances at different 'trigger prices'. This system is used in the American RGGI system, which is also discussed in Chapter 4.

Using different free allocation formats such as benchmarking in the absence of an emissions cap will, by contrast, allow industries to grow and prosper while limiting the cost burden on producers. The dilemma that policy makers face, of course, is that technological environmental protection standards will only temporarily limit emissions. Economic growth will catch up and give rise to increasing emissions again. This thus undermines the environmental effectiveness of the system.

Offsets

Economic growth could also be facilitated by granting the emissions trading system better access to cheap emission reduction allowances from abroad: known as offsets. By way of such offsets the environmental integrity of the emissions trading scheme can be ensured provided, of course, that emissions abatement has indeed to be realized (that is, that the quality of the offsets is ensured). Cheap offsets provide for a lower cost burden on production and can thus have a positive stimulus on the economy. The drawback, however, is that it can lower the market price of emission allowances to such an extent that it is too low to provide adequate incentives for technological innovation. This is one of the reasons for the current oversupply situation in the EU ETS, which is discussed further in Chapter 5.

Support schemes

Another means of allowing economic growth is to support hard-pressed industries. This support to 'energy-intensive trade-exposed' (EITE) industries can take the form of free allocation or direct subsidies (for example, state aid). This industrial policy and the granting of aid is generally seen as negative from an economic efficiency perspective since it is difficult to identify whether an industry prospers because of or in spite of the support received. Once such support measures are granted, politically they are very difficult to take away. It may also be feared that other countries would introduce similar protectionist measures to support their own industries. Despite the efficiency concerns, such measures are a means to allow those industries to grow – at the expense of other industries and non-covered sectors, which will need to make a stronger effort to abate greenhouse gases. Most emissions trading schemes provide for special support measures for EITE industries. Political support for such measures therefore appears to be available in most countries. Yet political support may change once the costs of these measures begin to mount and are clearly felt by those that have to bear them.

Intensity targets

An alternative method of allowing for economic growth is by way of an intensity target. Intensity targets require the reduction of carbon emissions per unit of GDP or per unit of production: China sets its intensity targets per unit of GDP. The benefit of this system is that emissions can increase with economic activity, but the economy undergoes a transition towards carbon efficiency that requires technological innovation and investment. This type of emissions trading system requires a sufficient fine-tuning of the target as emissions or GDP varies over time, and the stimulating of investment in abatement technology. The related policy objectives of reducing emissions and stimulating investment require: (i) the existence of a continuous price signal that induces investment in clean carbon technology; and (ii) direct influence over the amount of GHG emission allowances in circulation, in order to account for different levels of economic growth. This requires a system that combines both price controls and quantity controls. One way of approaching this dual constraint is to combine taxation with emissions trading. Depending on the particular design of the system, such a combination can give rise to double payments and high administrative costs, and can undermine marginal abatement costs.[20] There is, however, another design choice that

[20] For a full treatment of the points made here see Weishaar and Tiche (2013).

will allow policy makers to influence both the quantity of available allowances and the market price: reserve price auctioning. This involves the application of an auction reserve price which permits governments to directly determine the supply and the price of emission allowances on the market. Allowances not sold at the reserve price must be cancelled to ensure that the market price is not deflated by the expectation of the market participants that allowances will come onto the market before the end of the trading period. Reserve price auctioning constitutes an efficient form of achieving intensity targets while avoiding the drawbacks of double payments, high administrative costs and undermined marginal abatement costs.

3.5 Investment Incentives

Incentivizing investment in clean technologies is an important objective and is often presumed to stem directly from the scarcity of emission allowances within an emissions trading system. In cases of over-allocation or oversupply as a result of an economic downturn, market prices for allowances may decline significantly. As a result of price volatility, entrepreneurs shy away from long-term investment since they cannot predict the business risks in such a regulatory environment. Investment in green technology therefore may not take place. The solution for stimulating investment is to ensure a sufficiently high and stable market price. There are several ways to ensure this. Approaches that could be used include quantity targeting (that is, increasing the scarcity of emission allowances), price floors, and auction reserve prices. Each will be discussed in turn.

One way to increase prices in an ETS is to increase the scarcity within the system. This can for example be achieved by (i) allocating fewer allowances; (ii) the government purchasing emission allowances to counter the prevailing oversupply; (iii) adding additional sectors; or (iv) linking with another emissions trading scheme that has a higher market price.

While the last two possibilities are straightforward and are probably less problematic from a policy perspective, the first two possibilities merit further attention. Allocating fewer allowances than originally foreseen will trigger ill-feeling on the part of covered entities. This will reduce trust in the system and trust in the government since a bargain made with the government no longer stands. Even though legally there may be no protection of legitimate expectations that could bar a government from changing the 'rules of the game', it may still be hard to muster political support for the change. The reduction in allowances will not only decrease the allowances at times when prices are low, but also when prices are high. This means that prices will be higher overall (particularly so as emission allowance markets are inelastic). Albeit effective, the result may be that the costs of emissions trading are higher than desired. Setting the optimal quantity of allowances to induce investment is inherently difficult since ETS designers will not know which allocated quantity of allowances will trigger the necessary price to incentivize investment.

Government purchases of emission allowances on the market will please those covered entities that are able to sell allowances since they will benefit from a higher market price. It will not please those firms that have to purchase allowances at a higher market price. In addition, taxpayers are unlikely to be pleased since they will ultimately pay the price of such a policy. Politicians could be criticized for 'managing' prices and for interfering with emissions trading markets. Such policies, therefore, may be difficult to implement in practice.

Another means of stimulating investment is to target the price. This can be done by placing a levy on every offset that enters the emissions trading scheme until a price, thought to be sufficiently high to incentivize investment, is reached. It was intended to implement such a system in Australia, but as a result of the unilateral linking of the system with the EU ETS these plans had to be abandoned. The measure will be effective in triggering investment provided there is a shortage of allowances on the domestic market and the levy for offsets is set sufficiently high. The optimal setting of an appropriate price level is inherently difficult since ETS designers will not be sufficiently familiar with the marginal abatement costs of covered entities.

A further possibility is to use auction reserve prices. In emissions trading systems that use auctions widely the government can set a reserve price below which allowances will not be sold. Unsold allowances should then be cancelled (that is, taken off the market) in order to create a more immediate price effect. A situation of a weak market price would be overcome, because the supply of new allowances would effectively be stopped. Reserve prices are thus similar to trigger prices below which 'oversupplied' markets will not be flooded with allowances. Reserve prices will, of course, work only if the reserve prices are set at a sufficiently high level to actually trigger investment. As in the case above, the problem of determining the right auction reserve price to incentivize investment in abatement technology is inherently difficult since ETS designers will not have the necessary information about the marginal abatement cost structures.

Political acceptability of the three options can differ significantly. The Australian emissions trading system with the levy component was negatively, and misleadingly, portrayed as a carbon tax. Similarly, the European Commission appears to take every effort to ensure that it is not associated with price regulation in the EU ETS, while it does not seem to be troubled by engaging in quantity regulation (which, of course, also has a very direct effect on the market price of emission allowances). Reserve prices in allowance auctions are incorporated in the Regional Greenhouse Gas Initiative in the US and are acceptable there. Since the reserve price allows the market to operate freely without any distortion above the trigger price, and even allows the market price to temporarily dive below the trigger price, it can be viewed as a 'milder form' of price regulation that still allows the use of market forces to incentivize green innovation.

3.6 Summary

This section has presented a number of policy objectives that policy makers may wish to achieve through an emissions trading system. Efficient solutions are available to achieve each objective individually. However, if policy makers wish to achieve more than one goal at the same time through one ETS, efficient solutions will be hard to find. Compromises will have to be made since the 'efficient solutions' differ from each other. It is therefore important that policy makers are very clear about what they wish to achieve before they start to consider the design features of a system. It must also be emphasized that, according to Nobel Prize Laureate Jan Tinbergen, for each and every policy target there needs to be at least one policy tool (the Tinbergen Rule).[21] This suggests that overcharging policy instruments with multiple goals is undesirable from an economic perspective.

4. DESIGN CHOICES

This section of the chapter presents a number of design options from which ETS designers can choose. It presents some of the variables that they can use to design a system and to calibrate it with the particular policy objectives. Of course, there are an infinite number of design options imaginable. Here we provide a comprehensive overview of the most important design building blocks. This enables ETS designers to

[21] See Knudson (2009).

approach the literature with the right questions in mind to ultimately design a system that meets their needs.

The issues examined below include: (1) participation and coverage; (2) target setting; (3) cap-and-trade, credit trading and intensity targets; (4) monitoring, reporting and verification; (5) energy-intensive trade-exposed industries; (6) allocation rules; (7) closure and transfer rules, and *ex post* adjustment; (8) linking; (9) flexibility measures; and (10) the secondary market.

4.1 Participation and Coverage

Emissions trading systems can rely on voluntary participation, or participation can be mandatory. Voluntary schemes will attract those enterprises that would stand to benefit from an emissions trading system. There will thus be a selection bias in the sense that only those enterprises with low abatement costs will take part. Yet it is the heterogeneity of abatement costs that allows an emissions trading system to deliver cost-effective abatement. In the same vein, the environmental effectiveness of a voluntary scheme may be undermined if it is predominantly 'green' enterprises, with relatively less abatement potential, that take part.[22] In order to incentivize the participation of a wider range of enterprises, associated policies (carbon taxes, for example) could be applied to non-covered sectors. Covered enterprises could then be exempted from those measures.[23]

Another reason why mandatory schemes appear to be more efficient than voluntary schemes is that a larger number of participants yields more market liquidity and hence lower abatement costs.[24]

Opt-in and opt-out rules allow some flexibility for enterprises to choose between different policy instruments that pursue the same objective. Such ETS design features could be efficient since policy makers may lack the required information on the abatement costs of particular operators. This may lead some firms to leave the ETS, while others may join. This may lower the abatement costs that could be achieved via an emissions trading scheme and also enhance the political acceptability of the system. The administrative costs of such opt-in and opt-out rules,

---

22 See Zwingmann (2007) p. 98.

23 This is the case in the Swiss ETS.

24 Although it must be observed that the benefit of increasing the coverage of an emissions trading system finds its natural limit when the transaction costs associated with the inclusion of additional companies exceed the efficiency gains of the system: see OECD (1999) p. 21.

however, are higher than would be the case without these special derogations. In Switzerland, for example, enterprises with more than 20 MW total heat capacity are subject to the Swiss ETS. Operators with between 10 MW and 20 MW of heat capacity are allowed to opt in and take part in the ETS, and are thereby exempted from paying the Swiss carbon tax. Certain participating enterprises may also opt out of the Swiss ETS: that is, enterprises that have surpassed the 20 MW threshold but have emitted less than 25,000 tonnes of CO_2 in the past three consecutive years. They will then be required to pay the carbon tax. With these measures the legislator allows operators at the threshold level some flexibility to select the most favourable scheme for themselves.

Closely related to the question of who should be included in an emissions trading scheme is the question of which emissions should be included. Not only is it important to decide upon the types of emission to be covered (all six Kyoto Protocol greenhouse gases, or only a subset of them?), but also whether the system should target emissions solely from (fuel) inputs or include process emissions. Covering both types of emission allows for a more accurate internalization of externalities but generates higher monitoring costs. Policy designers will therefore need to balance the benefits of capturing process emissions and having more abatement possibilities against higher administration and operating costs.

While emissions trading systems mainly extend to enterprises, it is commonly recognized that households and individuals also contribute to global warming. There is a small but growing awareness by environmentally minded citizens who seek to reduce their personal 'carbon footprint' by covering their own carbon costs from, for example, travel activities. Citizens can conveniently calculate the carbon emissions of their last holiday flight online and pay the associated carbon costs for purchasing the required amount of emission allowances. If ETS designers are sympathetic to such activities, they may extend access to the trading scheme to non-covered entities.

Emissions trading is usually applied to large emitters, such as power companies and steelworks, but in principle it could be applied also to small emitters. Large emitters are likely to reap higher costs savings from emissions trading compared to the administrative costs they will bear of setting up and working with the system. Small emitters, such as hospitals or universities, will probably have small gains from emissions trading compared to the administrative costs of building and running a system of tradable emission rights. Therefore, a well considered trade-off must be made between the efficiency gains from emissions trading and the administration costs of including small emitters in the trading system.

An emissions trading scheme could also be extended to individuals. Although there has been some research, a few legal proposals and one or two experiments, no personal carbon trading scheme has yet been implemented anywhere in the world. In the context of climate change, for instance, personal carbon trading would imply that each adult must acquire an additional bank account: a carbon bank account.[25] Individuals receive a limited number of emission allowances each year, which they can administer. They must submit those allowances when buying climate-unfriendly products covered by the scheme – for instance, when a car driver purchases fuel at a service station or a household pays its yearly gas and electricity bill. Carbon 'chip and pin' cards could be used in shops and outlets such as fuel stations. Climate-unfriendly activities would reduce the number of allowances, but if an individual were to move to a smaller and better insulated apartment, buy a 'cleaner' car or purchase a more energy-efficient boiler, he will not just save energy but will also use fewer allowances. If this climate-friendly behaviour is strong enough, he may even save allowances which he can sell on the carbon market. This would also work the other way around if he wishes to have more allowances: for instance, if he buys a second car or moves to a much larger house, he will have to buy additional allowances to cover the emissions increase. This is not only efficient as it incentivizes climate-friendly behaviour, but it is also fair in the sense that the relatively big polluters pay more and those who emit less pay less.

This sounds attractive and promising, but there are, unfortunately, many design and implementation problems, some of which are set out below. While some issues relate to the operation of the system, others relate to distributional choices. Should every adult receive an equal number of allowances? Should adults receive more allowances if they have children living at home? Should people in rural areas receive more allowances if they have to travel long distances to work? How about tourists? Can individuals handle a carbon bank account on their computer? Do people want to think about buying, selling or banking emission allowances, especially since the carbon price is not fixed but inherently uncertain? Should the allowance price be capped, which could raise acceptance but reduce efficiency? Should the government monitor the emissions of each and every individual or is there a kind of 'smart' design possible to reduce the associated administrative costs? How high should the penalty be for those who do not comply with their emissions limits? Could the legal system efficiently handle carbon debtors?

---

25 Woerdman and Bolderdijk (2010).

The issues raised above should provide food for thought for emissions trading designers considering such a scheme.[26] We do wish to emphasize that several of these issues can be addressed. Administration costs could be reduced by (i) allocating free allowances 'downstream', based on the average fuel use per adult person in a reference year; (ii) using chipcard technology for car drivers; and (iii) concentrating monitoring and enforcement 'upstream' on fossil fuel producers and importers.[27]

First, a simple, generic allowance allocation rule reduces administrative costs, although it could lead to political opposition from individuals with higher than average emissions. Second, chipcards would need to be produced and sent out, although the cost of these cards is not very high. Third, it is not necessary to monitor each and every individual. For every unit of fossil fuel bought from a distributor, the individual would have to hand over a corresponding number of allowances. Distributors, in turn, could only obtain fuels from their suppliers in exchange for allowances. In this way, all allowances end up in the hands of the producers and importers of fuel, who would then be required to turn over the allowances to the environmental authorities for the carbon contained in the fossil fuels they had sold. As a consequence, allowance allocation would occur downstream at the level of individuals, but monitoring of emissions (via fuel sales) and checking whether they match the allowances would occur upstream at the level of producers and importers of fuel, whose number is usually limited.

Arguably, personal carbon trading is most easy to implement in countries where citizens and politicians are already used to emissions trading for large emitters. Examples are the European Union countries with their emissions trading scheme. However, it is precisely in those countries where perhaps the greatest problem in implementing personal carbon trading can be found: the problem of double counting. People already pay for their carbon emissions in their electricity bill, since power companies pass on a large element of the carbon price to their customers. If a personal carbon trading scheme were to be imposed on energy consumers, they would pay twice for the same emissions. This is likely to make a personal carbon trading scheme unacceptable for those consumers, as well as for politicians.

One of the advantages of personal carbon trading is that the emissions of individuals will be capped and individuals will therefore receive a more direct and more visible carbon incentive compared with existing

---

[26] See also Fawcett and Parag (2010); Starkey (2012).

[27] Woerdman and Bolderdijk (2010).

policies. These effectiveness and efficiency properties are likely to have a positive impact on energy conservation behaviour. Research suggests that the law, via instrument choice and institutional design, is able to change the 'environmental mind'.[28] However, the disadvantages of personal carbon trading include double counting of emissions, allowance price uncertainty, and the perceived (un)fairness of any particular allowance allocation rule. In addition, there are legal ramifications that would need to be considered when seeking to restrict consumption by individuals.[29]

In any event, it is important to note that there are other design variants that could give individuals a more prominent place in emissions trading than the possibility just sketched above.[30] An additional option, which would broaden the scheme but would also increase its complexity, is to include offsets in the trading scheme for individuals. This could help to change food-buying habits, stimulate green energy production and prevent deforestation.[31]

Building upon this literature, it is interesting to observe that personal carbon trading is cautiously and slowly entering the policy arena. In the United Kingdom, for instance, the Environmental Audit Committee of the House of Commons has studied the issue and argued in favour of personal carbon trading, but also drew negative conclusions as to its acceptance. Acceptance was less of a problem in Australia where Norfolk Island, with approximately 2,500 inhabitants, held a full-blown three-year test site for trying out a trading scheme for reducing individual emissions.[32] Although such cautious beginnings are to be welcomed in investigating the way in which consumers act upon such a scheme, establishing a full-blown personal carbon trading system from a policy perspective seems to be out of the question for years to come.

4.2 Target Setting

Setting the objective of an emissions trading system is not a trivial matter, and will take place within the context of significant uncertainty. Policy makers could set an absolute number of permissible emissions or relative targets per unit of production. They could also decide to set relative emission targets per unit of GDP. The nature of the target will

---

[28] Feldman and Perez (2009).

[29] De Cendra de Larragán (2013) p. 35.

[30] See, e.g., Fawcett and Parag (2010); Starkey (2012).

[31] Roy and Woerdman (2012).

[32] See http://www.norfolkislandcarbonhealthevaluation.com (last accessed 25 May 2013).

greatly impact on the design choice. For designs that address solely the quantity of emissions, a cap-and-trade system is presented below. For systems seeking to ensure relative targets, a credit-and-trade system is considered. There are then other designs that are suited to achieve targets that require both regulation of prices as well as the quantity of allowances.[33]

ETS designers should be aware that 'Rome was not built on one day', and must understand that a stringent emissions trading system that requires strong abatement in too short a period of time will not be efficient. As an emissions trading system, it will be able to produce cost-efficient abatement, but it will not be efficient from an intertemporal perspective. This will be the case, for example, when covered entities are forced to invest in the currently available technology without being given the time to develop cheaper and more efficient means of abatement. Limited resources may be spent on less efficient technology and create undesirable path dependencies.[34] To mitigate this risk, as well as to limit the associated adaptation costs of the economy, it would appear to be useful to increase the stringency of emission abatement targets over time. These insights are observed in both the European Union and Swiss trading systems, both of which use an annual reduction factor of 1.74 per cent. This means that the number of allowances that are available every year is decreasing linearly by that amount.

Designers can choose various ways of increasing abatement efforts over time. They could linearly (or exponentially, or in a step-wise fashion) reduce the number of available allowances. They could also choose to influence the market price of allowances, to set time limits for the validity of the available emission allowances,[35] or to have the nominal abatement value of the allowances debase over time. Creating more scarcity is a desirable feature of an ETS system. If relative targets are used, ETS designers can, in addition, set the level of future production standards so as to inform industry about the shape of the regulatory regime and to incentivize investment. Covered enterprises must be well informed of the future developments of the system in order to incorporate that in their investment decision-making. Predictability and credibility is crucial to trigger behavioural incentives for covered enterprises.

---

[33] On this issue see Weishaar and Tiche (2013).

[34] See Zwingmann (2007) p. 108.

[35] Emissions trading allowances can be valid for one or several trading phases.

4.3 Cap-and-Trade, Credit Trading and Intensity Targets

One of the most important decisions to be taken by an ETS designer is to determine whether the system should be based on a cap-and-trade system or on a credit-and-trade system. These are typically considered to be the two most widely used emissions trading variants. A cap-and-trade system fixes the maximum amount of allowances that can be emitted in the scheme. A credit-and-trade system, by contrast, has no emissions cap. Here, emissions rights are determined on the basis of over-compliance or under-compliance with production standards set by the government. With the application of emissions intensity standards per unit of GDP (rather than per unit of production) in China, one perhaps needs to add 'intensity trading' as a third variant. All three will be briefly introduced.

Cap-and-trade

A cap-and-trade system is an emissions trading scheme in which the overall amount of emission allowances made available by the government to covered entities is subject to a quantity limit, called a 'cap'. Under this system there is thus a maximum emissions ceiling installed by the government, but such schemes often permit additional emission allowances to be imported into the scheme (referred to as 'offsets', discussed below). Each participating enterprise covered by this type of system has to submit at the end of a time period (normally one year) enough greenhouse gas allowances to comply with its emissions.

Credit-and-trade

In a credit-and-trade system, emitting entities are free to produce but have to compare their actual emissions level per unit of output with that prescribed by a government standard. Credit-and-trade systems are also referred to as performance standard rate (PSR) systems in the literature. The emitting entity has to account for every emitted tonne in excess of this standard and pay a penalty if it cannot buy greenhouse gas emission allowances via the emissions trading system or benefit from past or future savings.

One important feature of the PSR trading system, which distinguishes it further from ordinary relative standard-based allocation systems, is that the emission allowances are not created by the government and distributed to emitting entities. Legislation obliges emitting entities to meet particular emissions targets, to have them verified by third parties and to report them to the government. While the government takes an active role in monitoring compliance with the law, it does not participate as an actor. Government takes a *laissez faire* approach, limiting itself to creating the rules to be followed by emitting entities. In this sense, the PSR system is operated by private entities which are able to sell their accredited savings to other market participants via a cost-effective trading system.[36]

It has been suggested for credit-and-trade systems that they had high transaction costs because they required each credit generation to be approved by a regulatory authority.[37] The economic effect of various transaction costs associated with different allocation mechanisms is to limit the degree of allocative efficiency by reducing the range over which mutually beneficial exchanges are possible.[38] A PSR system must therefore be designed to minimize those costs by relying on clearly defined standards that can be easily monitored, reported and verified by (possibly) centralized authorities.

Intensity standards/targets

In the Chinese 12th Five-Year Plan (FYP) (2011–15) the country has set itself a number of targets: (i) to reduce energy consumption per unit of GDP by 16 per cent; (ii) to raise the proportion of non-fossil fuels in the primary energy mix to 11.4 per cent; and (iii) to reduce by 17 per cent the carbon intensity per unit of GDP below 2005 levels by 2015.[39] All of these targets are set in relation to GDP rather than greenhouse gas emissions (GHG). This means that the target of the emissions trading pilots that are based upon this five-year plan are targeting only GHG emissions in conjunction with GDP. Thus emissions trading systems based upon such intensity targets consider two variables simultaneously. If, for example, the financial sector prospers, industry may emit more. If, for example, as a result of a financial crisis the GDP declines, industry will need to abate emissions even if the real economy (that is, that part of the economy that is actually producing goods and services) is not otherwise affected. This also means that, when deciding an intensity target, a significant amount of information is required on the economy in general and the covered industries in particular.

If, shortly before 2015, the economic growth in China turns out to be lower than expected, meeting the intensity target could be realized only if covered entities quickly intensify their abatement efforts. This, in turn requires the ETS designers to be able to set strong behavioural incentives to lead market participants, ideally by retaining influence over both the

---

36 This passage draws from European Commission (2003) p. 13.

37 Tietenberg, Grubb, Michaelowa, Swift and Zhang (1999) pp. 26, 34 and 107.

38 See Harrison and Radov (2002) p. 52.

39 See Sandbag (2012) p. 12; NDRC (2012).

allowance price as well as the allowance quantity. While such intensity targets allow for economic growth in a prospering economy, they may result in significant costs to covered enterprises since they will not be able to predict their investment risks. Investments that appeared profitable could turn out to be obsolete as a result of a sudden and unexpected sharpening of the emission allowance market, hence generating costs comparable to 'standards costs'.[40] In addition, it may require entities to reduce emissions unexpectedly and quickly, and thereby require them to invest heavily in existing technology without being able to develop more efficient, and hence less costly, means of abatement.

4.4 Monitoring, Reporting and Verification

Monitoring, reporting and verification (MRV) is at the heart of any emissions trading system. Correct data is needed for a well-functioning scheme and therefore deserves much attention. The associated administrative costs will be dependent upon the number, size and complexity of the installations and the covered processes. ETS designers must therefore consider the trade-off between accuracy in measuring additional emissions and lower abatement costs, on the one hand, and higher administrative costs on the other. The particular design of these rules is, of course, among other factors, dependent on the installations that will be covered and cannot be easily described in abstract. Chapter 7 of this book addresses critical MRV issues.

4.5 Energy-Intensive Trade-Exposed Industries

Energy-intensive trade-exposed (EITE) enterprises are considered to be vulnerable to price increases in their production costs stemming directly from CO_2 prices or indirectly from increasing energy prices. It is not surprising, therefore, that most, if not all, trading schemes have special rules to mitigate the exposure of the affected enterprises. Policy measures that are frequently chosen include subsidies, free allocation of emission allowances and price support schemes (for example, selling additional emission allowances if prices rise above a threshold level). While political acceptability of emissions trading may often hinge on the support of those disadvantaged sectors, it should be mentioned that

[40] Existing investments which are rendered unprofitable as a result of the introduction of environmental regulations are commonly referred to as 'stranded costs'.

carbon leakage is often overrated.[41] Support schemes are also inherently positional in the sense that if two countries support their cement industry, one could end up with a situation where neither becomes more competitive vis-à-vis the other.[42] As is generally feared to be the case for protectionist trade measures, if one country introduces support schemes, other countries could do the same. Both countries would be better off in not having support schemes in the first place – at least if their major trading partners were also to adopt similar measures to internalize environmental costs.

Setting up support schemes for EITE industries may also mean that other covered sectors would need to abate more. This would be the case if support for the EITE industries incentivizes production increases and thereby distorts marginal abatement costs. If there is a national target then perhaps even non-covered sectors, which are assumed to have high abatement costs, may need to contribute more to greenhouse gas abatement. To the extent that marginal abatement costs differ, this would be inefficient.

4.6 Allocation Rules

Allocation rules are needed to bring emission allowances onto the market. They therefore deal with the initial distribution of emission allowances among market participants, but not with the operation of the emissions trading market – the secondary market. There are several design choices from which policy makers can choose.[43] Auctions and other financial allocation systems, for example, require market participants to pay for their emission allowances. The difference between the two is that auctions allow market participants to determine the price they will pay in accordance with the prescribed auction rules, while other financial administrative allocation mechanisms allow ETS designers more discretion in price determination and final distribution. In contrast with the two preceding mechanisms, free allocation mechanisms do not require payment for allowances. System designers can base their allocation decision on historic emissions data (referred to as 'grandfathering'), on relative production standards, or any other administrative criteria. In

---

[41] See, e.g., Mattoo, Subramanian, van der Mensbrugghe and He (2009); Heilmayr and Bradbury (2011).

[42] They would, however, become more competitive vis-à-vis third countries that do not use such support measures.

[43] For a typology of allocation mechanisms in a static and dynamic context see, e.g., Weishaar (2007a).

contrast with the direct distribution of allowances to individual operators, allocation under relative standards is directly dependent on the emissions intensity of production. Operators complying with set emissions standards receive the right to produce and may be awarded credits which they can sell on the market if they exceed compliance with the standards.

The various allocation mechanisms will give rise to different behavioural incentives for emitters. The various design possibilities will also have an impact on the competitive position of enterprises.[44] Naturally, different distributional consequences arising from the allocation of allowances will affect the political support for the emissions trading system.

4.7 Closure Rules, Transfer Rules, *Ex Post* Adjustment

In the context of designing rules for the allocation of emission allowances, special rules must be created for situations when installations close down, reduce their production or shift their production to more efficient installations. Closure rules address the question of whether appropriate behavioural incentives are in place for entrepreneurs to close environmentally inefficient installations. Transfer rules determine whether there are appropriate behavioural incentives for entrepreneurs to shift production to the most environmentally efficient installations.

From an economic perspective it appears clear that incentivizing production by the most economically efficient facilities is desirable. Similarly, from an environmental perspective production should take place at the most environmentally efficient facility. The dilemma is that closure rules for installations and production transfer rules can produce incentives for entrepreneurs to continue to use a less environmentally efficient plant. When deciding to close one installation and transfer the production to a more efficient plant, entrepreneurs will take into account all costs. The less efficient plant will be closed. Yet, if by closing this less efficient plant the entrepreneur will no longer receive free emission allowances for that plant, he will take the value of those emission allowances into account when deciding whether to close or partially close the installation. Not allocating allowances for installations that cease production creates incentives to keep them open. This is inefficient. Similar inefficient incentives are set within the framework of the EU ETS in which allowances are not allocated in the year after the closure of the installation. This issue is discussed further in Chapter 5.

---

44 For a review of the effects on competition see Weishaar (2009).

When considering closure and transfer rules, distortions in competition also may arise. If the particular transfer rules differ between regions or emissions trading systems, an enterprise could close down in country A and (depending on the closure rules) continue to receive allowances there, and then reopen a production site in country B, where it would be eligible for allowance allocation under new entry rules. In such a case the operator of this production facility would receive more allowances than would its competitors, which would distort competition.

Discussions on closure and transfer rules are often discussed within the framework of '*ex post* adjustments', which are adaptations of the allocation rules after commencement of the operation of the emissions trading system. It may not be appreciated politically as it undermines the predictability of the system. One may wonder whether the impact of closure and transfer rules on the overall market liquidity would be sizable. One aspect of *ex post* adjustment that ETS designers should consider is the 'economic downturn'. Emissions trading systems work as an automatic stabilizer that alleviates the pressure on covered entities when they face economic hardship and the price of emission allowances is reduced. Low prices can affect investment in research and development. This issue is discussed further in Chapter 5 in the context of the current oversupply situation in the EU ETS.

4.8 Linking

Linking emissions trading systems allows market participants in one trading scheme to obtain emission rights issued in another trading scheme for domestic compliance obligations.[45] Linking schemes is expected to allow for a wider range of abatement opportunities and a larger market size, which stimulates liquidity and enables a more efficient resource allocation.[46] Since linking is presented in more detail in Chapter 9 of this book, it is not elaborated upon at this point. It is, however, important for ETS designers to consider linking when designing their schemes in order to prevent costly adaptation of the domestic system at a later point in time.

45 Haites (2003).

46 Baron and Philibert (2005).

4.9 Flexibility Measures

Emissions trading designers could decide to grant covered entities flexibility in terms of compliance. This could take the form of inter-temporal flexibility through banking and borrowing emission allowances, or in the form of territorial flexibility by allowing access to abatement opportunities abroad. This can be achieved by permitting covered entities to comply with their greenhouse gas emission obligations by using foreign emission allowances, referred to as 'offsets'. Granting such flexibility is considered to be efficiency enhancing, since it allows covered entities to take advantage of lower abatement costs elsewhere. Here, banking and borrowing will be discussed first; offsets will then be considered.

Banking and borrowing

In principle, emission rights are issued with a vintage year during which they are valid. Banking and borrowing provisions qualify this principle. Borrowing, on the one hand, allows covered entities to use allowances from future compliance years for the current compliance period. Banking, on the other hand, enables them to use allowances from previous vintage years in a future compliance year or period.

The economic intuition behind banking and borrowing provisions lies in the provision of temporal flexibility for regulated entities so that they can spread their abatement activities over time to cut their compliance costs. Because supply in a pure cap-and-trade system is inelastic, the ability of the supply curve to adjust over time depends on the ability to spread abatement over time.[47]

Fell et al.[48] categorize the motives for banking and borrowing into structural motives and motives related to uncertainty. Structural motives for banking and borrowing refer to the covered entities' need to reduce compliance costs through inter-temporal reallocation of abatement. If banking and borrowing are allowed, entities will redistribute their abatement across time and achieve abatement whenever it can be achieved at the lowest cost possible. The reasons for banking and borrowing related to uncertainty motives refer to the covered entities' need to reduce risk through inter-temporal reallocation of abatement. Emitters face all kinds of uncertainties relating, for instance, to (i) current or future costs of abatement; (ii) the stringency of future climate policies; (iii) the pace of

---

47 Fankhauser and Hepburn (2010).
48 Fell, Moore and Morgenstern (2011).

development of low-carbon technologies; and (iv) the possible integration of the emissions trading scheme under consideration into a global carbon market. Banking and borrowing allow covered entities to better manage such uncertainties.

The more flexible the banking and borrowing requirements are, the more cost-effective an emissions trading system is likely to be. In principle, banking and borrowing does not affect the environmental integrity of the trading scheme. However, this may not necessarily be true with regard to pollutants such as SO_2 emitters, where the amount of the environmental degradation in a given period depends on time-specific emissions.[49] Even with regard to stock pollutants such as CO_2 emitters, unlimited inter-temporal flexibility is not without danger. Unlimited borrowing, in particular, has several downsides. First, unconstrained borrowing may lead to the so-called 'moral hazard' problem by inducing firms to borrow from future compliance periods and exit from the market before achieving the time-shifted emissions. Second, a related problem (referred to as 'adverse selection') is that financially troubled firms are likely to borrow more than solvent firms. Third, regulated entities that borrow from future trading periods have an interest in softer future targets. In view of these political and economic challenges, most emissions trading schemes currently operating restrict the use of borrowing.

Unlimited banking between trading periods is allowed in a number of emissions trading schemes, such as:

- the EU ETS (except that banking Phase I allowances into Phase II was not possible);
- the Australian Carbon Pricing Mechanism (CPM) (except that fixed-price period allowances may not be banked);
- New Zealand; and
- the Regional Greenhouse Gas Initiative (RGGI) in the United States.

Borrowing is prohibited in all of the above emissions trading schemes, except for Australia, which only allows covered entities to borrow from future compliance periods to cover up to 5 per cent of their surrender obligations.

Moreover, it is important to note that borrowing is, in fact, possible in the EU ETS because of the timing in which allowance allocations are

[49] For a distinction between flow and stock pollutants see Lieb (2004). Flow pollutants cause an immediate impact on the environment while stock pollutants harm the environment only in the future as they build up.

made and compliance in respect of the previous year's emissions is required. Since allocations are made two months before regulated entities are required to surrender allowances covering their previous year's emissions, on any given compliance date (except the last year of a trading phase) such entities have two annual sets of European Union Allowances (EUAs) at their disposal. This allows them to 'borrow' EUAs from current year allocations to comply with their previous year's emissions. Inter-phase borrowing (borrowing from future trading phases) is, however, not possible.

Offsets

Carbon offsets refer to international emission allowances that are accepted into an emissions trading system for compliance purposes. They offer low cost abatement for covered entities and help to reduce the cost burden of environmental regulation on the domestic economy. Despite these attractive properties, there are also serious drawbacks. Some offsets are criticized for not leading to real emission reductions. Acceptance of such 'low quality' offsets would undermine the environmental effectiveness of the system. A further drawback is that, by allowing for offsets, the operators of the trading system are no longer fully in control of the supply of emission allowances. Not only the market supply – that is, the quantity of abatement allowances available for compliance – is more difficult to determine, but also the market price. Offsets tend to be cheaper than domestic abatement opportunities and may lead to a lower than expected allowance price. This is attractive, but may not be appropriate if a market price is desired to incentivize technological innovation.

Various types of offset can be made available to covered entities. Here, the most prominent terms are described.

- Removal units (RMUs) are allowances generated by an improved national carbon sink performance. They are issued by Kyoto Protocol Annex I Parties for carbon absorption by land use, land-use change and forestry (LULUCF) activities.
- Emission reduction units (ERUs) stem from joint implementation (JI) projects that are carried out in industrialized countries.
- Certified emission reductions (CERs) stem from Clean Development Mechanism (CDM) projects in developing countries.
- Temporary certified emission reductions (tCER) and long-term certified emission reductions (lCERs) are emission reductions stemming from afforestation and reforestation CDM projects. Temporary CERs expire at the end of the following commitment period,

but may be renewed if CO_2 sequestration can be proven. Long-term CERs, by contrast, expire only when the project itself ends.

In addition to the offsets mentioned above ETS designers can allow for the use of assigned amount units (AAUs). These are emissions allowances assigned to members of the Kyoto Protocol during a particular commitment period.

4.10 The Secondary Market

Secondary markets are important to allow for cost-effective exchanges of emission allowances. The lower the transaction costs associated with sales and purchases of allowances and the lower the overall abatement costs, the higher will be the costs savings of the emissions trading system. Finding a trading partner for bilateral exchanges is associated with high search costs. More efficiently, brokers may be used if permitted in order to facilitate exchanges. Similarly, exchanges can be used to allow for cost-efficient transactions. When designing secondary markets, due account must be paid to market oversight, prevention of fraud and (cyber) crime, money laundering, market manipulation and market abuse. Chapter 6 addresses sensitive issues in relation to the secondary market; they will therefore not be elaborated upon here.

5. CONCLUSION

This chapter has presented several design variants of emissions trading systems. The choice not to present the material along the lines of academic typologies was deliberate. Taking a more practical approach, focusing on the wider framework within which an emissions trading system is to operate, discussing policy objectives and the building blocks available for designers to construct an ETS, seems to be more useful. In that way ETS designers have a better understanding of relevant issues so as to enable them to approach the academic literature with the right questions in mind. Instead of offering clear answers to designers, we hope to have provided a better idea of the relevant questions to be asked and trade-offs to be made.

As a consequence, the insights of this chapter are not geared to present the 'perfect' emissions trading system, but rather to foster the understanding that an ETS can only be 'perfect' in its natural habitat and given the policy objectives that it is supposed to address. There is thus no 'one size fits all' system that can be acquired by way of legal transplant – copying and pasting the design of existing systems from abroad or from other environmental policy fields.

Despite our understanding that policy makers seek to achieve different aims at the same time, we wish to stress that it may not be desirable to charge a system with too many goals. Integrating more policy goals will make emissions trading systems more complex and complicated and may thus negatively influence their efficiency and performance.

4. Real-life applications of emissions trading systems

1. INTRODUCTION

This chapter presents an overview of emissions trading systems that are currently operating. In spite of some imperfections in and criticisms of emissions trading, there is a quick proliferation of such systems around the globe. An effort has been made to give a concise presentation in this chapter of prominent schemes.

The reader should be aware that the systems presented below differ considerably with regard to their size, scope and coverage, and their design. While some systems, such as the European Union Emissions Trading Scheme (EU ETS), are supranational in nature, others are sub-national initiatives, such as the American RGGI and the WCI.[1] Yet other schemes are national or provincial (such as those in Australia and Guangdong), or restricted to city boundaries or metropolitan areas (for example, the Tokyo and Shanghai systems). While their respective size and ambitions naturally diverge, so does their potential for greenhouse gas (GHG) reduction. It appears that the city-based emissions trading schemes focus primarily on carbon dioxide (CO_2) emissions arising from fuel inputs and energy consumption, while larger schemes are able also to incorporate process emissions.

The emissions trading systems in Europe, North America, and the Asia-Pacific region are discussed first, followed by the Asian systems. Issues such as policy targets, start dates and compliance dates, allocation mechanisms, coverage, permissible types of allowance and offset, support schemes/leakage safeguards, fines for non-compliance, banking and borrowing, and linking will be considered. Given the brevity of this chapter and the number of systems considered, it is not our ambition to be complete in every detail. Our aim is to describe current systems and to show that different jurisdictions have chosen different ETS designs.

---

[1] The Regional Greenhouse Gas Initiative and Western Climate Initiative, both operating in the United States.

2. EUROPE

2.1 European Union Emissions Trading System

The Council of the European Union[2] adopted a climate change action and renewable energy package in April 2009 to achieve a 20 per cent reduction in greenhouse gases (below 1990 levels) and a 20 per cent share of renewable energy in the total energy consumption by 2020, and to meet its targets of the Europe 2020 Strategy of smart, sustainable and inclusive growth.[3] The climate change action and renewable energy legislative package represents a multi-faceted approach, supporting the attainment of Europe's strategic objectives[4] and to support each Member State in reaching its respective targets. The energy efficiency elements contained in the 2020 Strategy are further targeted by the Energy Efficiency Plan 2011,[5] which proposes ways to strengthen European efforts to meet its target of a 20 per cent energy efficiency improvement below the 2020 predicted levels by that year. Energy efficiency addresses energy generation, transformation, distribution and consumption. It also seeks to foster a large array of incentives and behavioural changes to stimulate building renovation, to green public procurement rules, to incentivize building improvements, to encourage proliferation of best practices and to have more energy audits. The current cap of the European ETS reduces over time so that in 2020 the emissions from covered sectors will be 21 per cent lower than those in 2005. Depending on international action, the European Union is prepared to sharpen its goals to 30 per cent below 1990 levels by 2020. For 2050 it envisages reductions of 80 to 95 per cent below 1990 levels.

With the current 2020 goals for smart sustainable and inclusive growth, the EU is addressing all areas necessary to mitigate global warming: it leads to a reduction in allowances per person (energy efficiency), it addresses the emissions intensity of the economy (EU ETS) and carbon-friendly energy production (renewable energy). While addressing all areas contributing to climate change makes intuitive sense, it must be observed that the three policy areas affect each other. More renewable

---

[2] See Council of the European Union (2009).

[3] See COM(2008) 30 final and COM(2010) 2020 final.

[4] The EU is not projected to meet its Europe 2020 strategy objective of enhancing national energy efficiency by 20% in 2020 and additional measures are proposed by the Commission: COM(2011) 370 final.

[5] COM(2011) 109 final.

energy means that less energy production takes place within the framework of the EU ETS, which in turn reduces the market price for emission allowances. This gives rise to a 'waterbed' effect in which advances in one policy area lead to less pressure in another policy area. Similarly, if private homes need less heating, energy demand is lower and hence there will be less demand for emission allowances by energy producers covered by the EU ETS. Since the scheme is designed as a cap-and-trade system, it also means that other covered sectors will have fewer incentives to invest in abatement technology. Reductions in renewable energy and energy efficiency are thus partially outweighed by additional reductions in the industry sector. This interaction effect is, however, not a one-way street. It works both ways, since higher emission allowance prices incentivize investment in energy efficiency and renewable energy.

The EU ETS, a cap-and-trade system for GHG emission allowances for energy-intensive installations, can rightly be called the cornerstone of the European Union climate change policy. It extends to all 28 Member States of the EU, as well as Norway, Liechtenstein and Iceland, and covers around 45 per cent of European greenhouse gas emissions. The legal framework was set up in 2003,[6] obliging around 5,000 operators with more than 11,000 installations to participate in this multi-jurisdictional regulatory system to reduce CO_2 emissions from four broad sectors: (i) energy (including electricity producers and oil refineries); (ii) ferrous metals (iron and steel); (iii) minerals (cement, glass, ceramics); and (iv) pulp and paper.[7] Aviation has since been added as a fifth sector. Gases actively covered include carbon dioxide, nitrous oxide and perfluorocarbons.[8] Installations with a thermal input exceeding 20 MW are covered by the EU ETS. Small installations may be excluded from the scheme if they emit less than 25,000 tonnes of carbon dioxide equivalent and where they carry out combustion activities that have a rated thermal input below 35 MW.[9]

The programme is implemented in multiple phases. The first phase was from 2005 to 2007; the second phase, which coincided with the Kyoto Protocol first compliance period, was from 2008 to 2012, the third period is from 2013 to 2020 and the fourth from 2021 to 2028. The first trading period (Phase 1) suffered from over-allocation of emission allowances and witnessed a dramatic fall in prices. From its forceful start at the beginning of 2005 prices increased, closing at above €27 per tonne in

---

[6] Directive 2003/87/EC.
[7] Ibid., Annex I.
[8] On the basis of Directive 2003/87/EC, Annex II (as amended).
[9] Directive 2009/29/EC, Art. 27.

July 2005. A subsequent fall could not shatter the trust of the market and prices, although more volatile than before, increased to above €30 per tonne in April 2006. The unannounced pre-release by several Member States of the 2005 verified emissions data in May 2006, and the lower than expected actual level of emissions, caused a drastic reduction in the European Union allowance (EUA) prices. Prices regained momentum and hovered at around €15 per tonne, but dropped after October 2006 to €0.10.

The second trading phase (Phase 2) appeared initially to operate more smoothly before, again, over-supply occurred as a result of the economic downturn. Allocation in Phases 1 and 2 was heavily biased in favour of free allocation (grandfathering, accounting for 95 and 90 per cent respectively) and marginally on auctioning.

In the second EU ETS trading period, Member States were subject to obligations under the Kyoto Protocol and under the EU Burden Sharing Agreement. From 2013 onwards, the Effort Sharing Decision applies, which means that Member States are required to develop national policies (and related regulatory measures) to address greenhouse gas emissions of sources not covered by the EU ETS.[10]

The third trading phase is characterized by a strong degree of harmonization both of free allowance allocations and of allowance auctions. Free allocations are made on the basis of *ex ante* benchmarks. The applicable formula is as follows:

> Benchmark × historic activity level × carbon leakage exposure factor × cross-sectoral correction factor or linear factor.

Benchmarks are product specific based on the top 10 per cent most efficient installations during the period 2007–08.[11] This benchmark remains valid until the end of the current trading phase. The historic activity level is calculated by either taking the average of the two median production values between 2005 and 2008 (thus excluding the highest and lowest value during this period) or by taking the average of the production levels for the period 2009–10. The higher of the two will be considered. The carbon leakage exposure factor is set at 1 for EITE industries and at 0.8 for non-EITE industries (the electricity sector is

---

[10] Decision No. 406/2009/EC of the European Parliament and of the Council of 23 April 2009 on the effort of Member States to reduce their greenhouse gas emissions to meet the Community's greenhouse gas emission reduction commitments up to 2020, pp. 136–48. The effort sharing decision covers all six Kyoto Protocol gases.

[11] Directive 2009/29/EC, Art. 10a(2).

generally excluded from free allocation). The latter will decrease over time to 0.3 by 2020; thereafter the intention is that it will be reduced to zero in 2027. Auctions are increasingly phased in as the predominant form of allocation, rising from 20 per cent (non-EITE sectors, excluding the electricity sector) in 2013 to 70 per cent in 2020, while 2027 is the targeted date for full auctioning.[12] In the aviation sector auctioning will be limited to 15 per cent during Phase 3. Auctioning is, with some exceptions for Eastern European EU members, the default allocation form for the power sector.[13] While 88 per cent of the total quantity of allowances to be auctioned will be distributed to Member States,[14] 10 per cent will be allocated for the purpose of 'solidarity' and growth within the EU.[15] The remaining 2 per cent will go to those Member States that were in 2005 at least 20 per cent below their Kyoto Protocol base-year emissions.[16] Fifty per cent of the auction revenues should (not shall) be used for emission-mitigation and adaptation measures.[17]

Emitting entities under the EU ETS are obliged to monitor and report on their emissions for each calendar year and have their emissions verified by an accredited entity. By 30 April of the following year emission allowances must be surrendered to cover all emissions. Failure to do so results in a fine of €100 per tonne of CO_2 and the obligation to surrender the missing allowances.

Sectors and sub-sectors exposed to a significant degree of carbon leakage are eligible to receive up to 100 per cent of allowances for free.[18] This can be reviewed when (or if) an international climate change agreement is reached.[19] With a view to achieving emission reductions and strengthening long-term regulatory certainty, the overall scarcity of emission allowances in the third trading phase and beyond will be achieved through a linear reduction factor of (currently) 1.74 per cent per year.[20]

---

12 Ibid., Art. 10a(11).

13 Ibid., Art. 10a(3). Note that some Member States were able to secure concessions to compensate for their poorly integrated electricity grid: see Directive 2009/28/EC, Art. 10c(1).

14 Directive 2009/29/EC, Art. 10(2)(a).

15 Ibid., Art. 10(2)(b).

16 Ibid., Art. 10(2)(c).

17 Ibid., Art. 10(3).

18 Ibid., Art. 10a(12).

19 Ibid., Art. 10a(1).

20 Ibid., Art. 9; it should be reviewed by 2025. Because of the current over-supply of allowances, some policy makers and academics plead in favour of strengthening this emission reduction factor.

In order to reduce the threat of carbon leakage, the Commission has adopted rules that permit Member States to adopt support measures for energy-intensive trade-exposed sectors that are subject to a significant risk of carbon leakage.[21] Eligible sectors include aluminium, copper, fertilizers, steel, paper, cotton, chemicals and plastics. Member States are allowed to grant subsidies of up to 85 per cent of the increase faced by the most efficient companies in the relevant sector during the period 2013–15, 80 per cent during 2016–18 and up to 75 per cent during the last two years of the third trading period.

Another means of mitigating the high cost burden of emissions trading is the use of offsets. In the EU ETS both Certified Emission Reduction units (CERs) and Emission Reduction Units (ERUs) are permissible. Since offsets are intended to be supplemental, there are quantitative restrictions on their use. These restrictions were determined at Member State level in the second period; from 2013 onwards they will be determined at the European level.[22] The overall use of permissible credits is not to exceed 50 per cent of the Union-wide reductions below 2005 levels of the existing sectors under the EU scheme over the period 2008–20, and 50 per cent below the 2005 levels of new sectors and aviation over the period from the date of their inclusion in the EU scheme to 2020. The European Commission has recently proposed more stringent rules for the use of offsets.[23]

Besides quantitative restrictions there are also qualitative restrictions. Offsets from nuclear facilities and forestry projects (LULUCF) are not permitted. Similarly offsets from large hydroelectric projects are not accepted unless they meet special requirements.[24] As of 1 January 2013 offsets from projects related to HFC-23 and nitrous oxides (N_2Os) from adipic acid production will no longer be permissible (certain transition rules apply).[25]

The use of offsets for the purpose of compliance in the second trading phase has demonstrated a marked increase.[26] The new qualitative restrictions would mean that 80 per cent of the offsets used for compliance in 2010 would no longer be admissible. The increased use of offsets during

---

21 Communication from the Commission (2012).

22 See Directive 2009/29/EC, Art. 11(a)8.

23 European Commission (2013).

24 See Directive 2004/101/EC, Art. 11(b).

25 See Commission Regulation (EU) No. 550/2011.

26 In 2008 the amount of offsets used for compliance was 82 million tons while in 2011 it was 252 million tons: See Commission Staff Working Document (2012).

the second period has contributed to the significant degree of over-supply that is undermining the emission allowance price in the EU.

Given the possibility of banking between the second and third trading period, the EUA price is still positive even though there is significant over-supply well beyond the year 2020. As a result of the rejection by the European Parliament of a proposed Decision to clarify provisions of Directive 2003/87/EC on the timing of auctions of greenhouse gas allowances,[27] the Commission's plans for time-shifting the auctioning of 900 million allowances to 2019 and 2020 (referred to as 'backloading') appears to have been stopped in its tracks.[28] This was mirrored in a large drop in the EUA market price. From one day to the next prices fell during April 2013 from €4.73 to €3.15.[29] The backloading proposal, however, is not yet off the table. A second round of voting on reducing more than 900 million tonnes of CO_2 allowances from the EU ETS is announced for a parliamentary committee for 19 July 2013. A second plenary vote in the European Parliament could also take place in July 2013.

2.2 Swiss Emissions Trading Scheme

Switzerland acceded to the Kyoto Protocol as an Annex I country and committed itself to reduce its greenhouse gas emissions by 8 per cent below 1990 levels by 2012. The Swiss target for the second Kyoto compliance period lies at 15.8 per cent below 1990 levels. The Swiss system operates on the basis of the CO_2 Act (CO_2 Gesetz),[30] which dates back to 1999 and has been revised recently to bring the Swiss ETS more in conformity with the EU ETS (an important stepping-stone towards linking both trading schemes). Greenhouse gas emissions will be reduced by 20 per cent below 1990 levels by 2020.[31]

---

[27] The motion to reject the Commission proposal COM(2012) 416 final was accepted by the European Parliament on 16 April 2013: see point 11 at http://www.europarl.europa.eu/sides/getDoc.do?pubRef=-%2f%2fEP%2f%2fNONSGML%2bPV%2b20130416%2bRES-VOT%2bDOC%2bPDF%2bV0%2f%2fEN.

[28] Draft Commission Regulation amending Regulation (EU) No. 1031/2010.

[29] The price of an EUA was €4.73 on 15 April 2013 and only €3.14 on 16 April 2013 when the European Parliament rejected the Commission Proposal. Price data taken from the European Energy Exchange, available at: www.eex.com (last accessed 22 April 2013).

[30] Bundesgesetz über die Reduktion der CO_2-Emissionen, 641.71vom 23. Dezember 2011 (Stand am 1. Januar 2013).

[31] See CO_2 Gesetz, Art. 3.

The CO_2 Gesetz and its implementation rules (CO_2 Verordnung)[32] were renewed in 2012 and entered into force on 1 January 2013. The legal changes bring the former Swiss emissions trading system into alignment with the EU ETS. The relatively small Swiss system will be able to link with the EU ETS and benefit from greater market liquidity and abatement possibilities in the EU ETS. The system covers carbon dioxide, methane, nitrous oxide, hydrofluorocarbons, perfluorocarbons, sulphur hexafluoride and nitrogen trifluoride.[33]

The Swiss Emissions Trading Scheme is a cap-and-trade system that covers installations with a total heat capacity (*Gesamtfeuerungswaerme-leistung*) of 20 MW or above. Those installations with a total heat capacity of between 10 MW and 20 MW may opt into the system. Entities taking part in the Swiss system will be reimbursed for the CO_2 taxes they have paid (*Abgage auf Brennstoffen*). Entities willing to opt into the system must declare this before 1 June 2013; once this opt-in declaration is made, it may not be revoked. Entities that surpass the 20 MW threshold, but have emitted less than 25,000 tonnes of CO_2 in the past three consecutive years, may declare an opt-out and do not have to take part in the emissions trading scheme. These entities must also make a declaration before 1 June 2013.

The third trading period of the Swiss ETS resembles that of the EU ETS, and will run from 2013 to 2020. On 30 April each year operators must surrender emission allowances to cover their actual emissions. Non-compliance with this obligation is subject to a fine of 125 Swiss Francs. As with the EU ETS, this fine does not release the defaulter from the obligation to surrender the missing allowances in the next compliance year.

The Swiss scheme is subject to an emissions cap. The overall cap is subject to an annual reduction factor of 1.74 per cent. Comparable with the EU ETS, the Swiss system employs free allocation on the basis of benchmarking and auctioning. The benchmarks and the allocation methodology closely resemble those employed by the EU ETS.

The actual allocation will be determined by multiplying the product-specific benchmark by the median of the historic production data of the period 2005–08, or alternatively 2009–10. The resulting product is multiplied with a correction factor that reflects the susceptibility for carbon leakage. For sectors that suffer from a significant risk of carbon

---

[32] Verordnung über die Reduktion der CO_2-Emissionen, 641.711vom 30. November 2012 (Stand am 1. Juni 2013).

[33] See CO_2 Verordnung, Art.1.

leakage, such as CO_2-intensive or trade-intensive sectors, this correction factor will be set at 1, which means that all allowances will be allocated for free. For sectors not exposed to risk of carbon leakage the correction factor will decrease linearly from 0.8 in 2013 to 0.3 in 2020. These operators must purchase the required emission allowances from other operators or at auction, or abate internally. The electricity sector is not eligible for free allocation.

If the installation produces goods that are not covered by a product-specific benchmark, then either a heat-benchmark or a fuel-benchmark (Brennstoff benchmark) will be used to determine the actual allowance allocation. If this is not possible, allocation will be based on historical emissions.

Offsets can be used to a limited extent, not exceeding 11 per cent of the allocation received during the trading period 2008–12. After 2013 permitted offsets are ERUs and CERs.[34] Temporary CERs and long-term CERs, or emissions reductions from voluntary markets, may not be used. The calculation of the maximum quantity of offsets that an undertaking is permitted to use depends on when it entered the Swiss scheme, but the differences in outcomes appear to be limited.[35]

3. NORTH AMERICA

3.1 United States Sulphur Dioxide Cap-and-Trade Programme

The United States (US) sulphur dioxide (SO_2) emissions trading scheme is the first ETS in the world. Following an amendment of the 1990 Clean Air Act, the SO_2 cap-and-trade system was launched in 1995. Its aim was to reduce SO_2 emissions from electricity production by 10 million tonnes below 1980 levels. The system was implemented in two phases. Phase I ran from 1995 to 2000 and ultimately affected 445 installations. Given the stated reduction objective, around 3.5 million tonnes of SO_2 emissions had to be reduced per year during Phase I. In Phase II, which commenced in 2000, the scheme was extended to include all new utility plants and plants with an output capacity exceeding 25 megawatts (MW), and set tighter emission limits. The scheme covers approximately 2,000 installations. An evaluation by the Organisation for Economic

---

34 Provided the quality criteria set out in the Ordinance on the Reduction of CO_2 Emissions are met.

35 See Schweizer Eidgenossenschaft (2013) pp. 19–20.

Co-operation and Development (OECD) showed that the instrument is both environmentally and cost-effective.[36]

An allowance represents the right to emit one tonne of SO_2. Phase II allocations are tighter than those in Phase I. Emission allowances are predominantly allocated for free on the basis of an installation's 'share in total fuel use of the industry in the late 1980s' and a specific emissions rate (grandfathering). Some allowances are also auctioned. The Environmental Protection Agency (EPA) is required to hold back approximately 2.8 per cent of allowances and to auction them every year. The auction proceeds are redistributed to affected entities.

Under the Acid Rain Program (of which the SO_2 emissions trading scheme is a component), each regulated unit is obliged to measure its SO_2, NOx and CO_2 emissions. Based on the obtained data, the EPA conducts an annual 'reconciliation', which is the process of comparing a unit's annual emissions and allowance holding. At the end of each year covered entities have a 60-day grace period to ensure that they hold sufficient allowances to cover their SO_2 emissions in the preceding year. Non-compliance with an entity's obligations leads to a fine of US$2,000 (adjusted for inflation) per tonne of SO_2 emitted in excess of its account holdings. The missing emissions must be covered in addition.

3.2 Regional Greenhouse Gas Initiative

The Regional Greenhouse Gas Initiative (RGGI – pronounced 'reggi') is one of several sub-national emissions trading systems in the United States. In 2005 nine north-eastern and mid-Atlantic states signed a Memorandum of Understanding (MoU) to create this trading system with scope to stabilize CO_2 emissions at 2009 levels by 2015 and to reduce emissions by 2.5 per cent per annum until 2018.

The system includes only CO_2 emissions from fossil fuel-fired power plants that are larger than 25 MW. Covering more than 200 electricity generators, it accounts for 95 per cent of CO_2 emissions from the electricity sector.

The aim of the system is to reach an emissions reduction of 10 per cent below 2009 levels by 2018. It is subdivided into three-year compliance periods, referred to as 'control periods'. The first control period was from 1 January 2009 to 31 December 2011 (Phase 1); the second from January 2012 until December 2014 (Phase 2), and the third will run from 2015 until the end of 2017 (Phase 3). At the end of each control period,

---

[36] Ellerman (2004) p 94.

emitting entities (referred to as 'CO_2 budget sources') must surrender eligible emissions allowances to cover all of their emissions in the relevant control period along with their compliance report. In the event of non-compliance, budget sources must submit three allowances for every ton (the unit of measurement adopted by the RGGI is the 'short' ton – 907.1847 kg) of CO_2 for which they have failed to submit.

Most of the allowances are sold in regional auctions and more than 80 per cent of the auction revenues are recycled back to consumers, into energy efficiency measures and renewable energy, direct energy bill assistance and other greenhouse gas reduction programmes. The auctions are subject to an auction floor price currently set at US$1.93 per ton of CO_2. The CO_2 allowance prices tend to be around US$2.30, and hence slightly above the auction reserve value.

The RGGI system does allow for offsets to prevent prices from rising. Permissible credits under RGGI must be generated from five eligible projects: (i) capture or destruction of methane from landfills; (ii) reductions in sulphur hexafluoride emissions from electricity distribution equipment; (iii) CO_2 sequestration through forestation; (iv) emission reductions through energy efficiency in buildings; and (v) methane reduction through agricultural management practices.

Up to a price of US$7 offsets are restricted to those from within the RGGI system and capped at a maximum of 3.3 per cent of outstanding RGGI allowances. Up to a price of US$10 eligible offsets may also be imported to the RGGI scheme from offset programmes within the United States. At prices above US$10 offsets may be imported up to a value of 10 per cent of outstanding RGGI allowances. These allowances are not restricted to the US, but can come from international offset projects and emissions units from foreign cap-and-trade schemes.

While carbon leakage does not (yet) appear to be a central concern, it was studied closely in the 2010–12 RGGI programme review and has been incorporated into the updated model rules that will need to be implemented by each participating state.[37] A cost containment reserve will be added to the scheme from which allowances can be made available if the allowance price reaches predetermined trigger prices. The cost containment reserve will be 10 million allowances per year and the trigger prices are set at $4 per ton in 2014, $6 in 2015, $8 in 2016, $10 in 2017, and thereafter will increase by 2.5 per cent annually. In addition, the RGGI states commit themselves to identify and evaluate potential

[37] For a summary of the proposed amendments to the Model Rule, see RGGI (2012).

import tracking tools and to assess the need for further potential leakage mitigation policy measures.[38]

3.3 Western Climate Initiative

The Western Climate Initiative (WCI) started as a collaboration between seven US states (Arizona, California, Montana, New Mexico, Oregon, Utah and Washington) and four Canadian provinces (British Columbia, Manitoba, Ontario and Quebec). Since 2007, WCI members have worked together to identify, evaluate and implement climate change policies. By December 2011, all US states except California had withdrawn from the initiative. The system thus consists now of four Canadian provinces and California.

The Western Climate Initiative is a cooperation of bilaterally linked regional cap-and-trade schemes that mutually recognize their emission allowances. The schemes aim to reduce emissions by 15 per cent below 2005 levels by 2020 and to create investment in and development of clean energy technologies and jobs and to protect public health. This objective represents the aggregate emissions reduction goal of the participants. The initiative was scheduled to start on 1 January 2012. Quebec and California are the first systems to have started their compliance periods, on 1 January 2013. Each cap-and-trade system will be based on the Design Plan that was commonly agreed upon in July 2010.[39]

Compliance periods last for three years and covered entities must surrender sufficient emission allowances by midnight on 30 June of the year following the end of each compliance period. Failure to do so results in a penalty of having to submit an additional three allowances for every tonne of CO_2 equivalent that was not submitted.

In the five WCI trading schemes emission allowances are allocated in accordance with regional rules. The WCI design agreement therefore does not prescribe which allocation methodology must be used. Benchmarking and auctioning (with a reserve price) are permissible. To meet the 2020 targets, the caps of the respective schemes are recommended to decrease linearly. *Ex post* adjustments have been ruled out except in those cases where there is a change in the scope of the schemes or WCI membership, or where there is evidence of incorrect or inaccurate cap setting.

---

[38] RGGI (2012) pp. 1 and 3.

[39] Western Climate Initiative (2010).

It will cover several greenhouse gases (carbon dioxide, methane, nitrous oxide, hydrofluorocarbons (HFCs), perfluorocarbons (PFCs), sulphur hexafluoride and nitrogen trifluoride) from the electricity sector (including electricity imported to the WCI), and industrial combustion at large sources, as well as industrial processes. As of January 2015 the system will also cover transportation fuels, and remaining residential, commercial and industrial fuels so as to include nearly 90 per cent of emissions.

The first two emissions trading systems that came into existence recently within the framework of the WCI – the Californian cap-and-trade system and the Quebec carbon market – are reviewed below.

Californian Cap-and-Trade System

The California Global Warming Solutions Act of 2006 (AB32) requires California to reduce its emissions to 1990 levels by 2020. The emissions for 2020 are set at 334.2 million tonnes of CO_2 equivalents, implying an increasing shortage of 3 to 4 per cent during the period 2015 to 2020.[40] For the year 2050, California strives to reduce its greenhouse gas emissions by 80 per cent below 1990 levels.[41] Emissions are covered from all six greenhouse gases regulated under the Kyoto Protocol (carbon dioxide, methane, nitrous oxide, HFCs, PFCs, sulphur hexafluoride) as well as nitrogen trifluoride and other fluoridated greenhouse gases.

The Californian system started operation on 1 January 2013 and is subdivided into different compliance periods. The first of these runs to the end of 2014; the second from 2015 to the end of 2017, and the last from 2018 to the end of 2020. Every year covered entities must surrender emission allowances equivalent to at least 30 per cent of their verified emissions.[42] The remainder of the outstanding emission allowances must be surrendered at the end of the compliance period.

The first compliance period covers around 35 per cent of California's GHG emissions and extends to all industrial sources emitting at least 25,000 metric tonnes of CO_2 equivalents and to electricity generated in or imported into California. During the second period, as of 2015 onwards, distribution of transport fuel, natural gas and other fuels will also be

---

[40] See § 95841, Table 6-1; Art. 5: California Cap on greenhouse gas emissions and market-based compliance mechanisms, Subchapter 10 Climate Change, Article 5, Sections 95800 to 96023, available at: http://www.arb.ca.gov/cc/capandtrade/september_2012_regulation.pdf (California cap on greenhouse gas emissions).

[41] See Office of the Governor, Executive order S-3-05 of 06 January 2005.

[42] See § 95855, California cap on greenhouse gas emissions, above n. 40.

covered by the emissions trading system, which will thus cover around 85 per cent of total emissions of the state. Approximately 350 large enterprises, representing about 600 installations, are covered.

The industrial sector will receive allowances on the basis of benchmarks (set at 90 per cent of average emissions), which will decline over time. The allocation for natural gas will be determined by the California Air Resources Board (CARB) before the beginning of the second compliance period. Allocation for electricity will be based on long-term procurement plans. The remainder of the allowances will be sold at auction. The auction is a sealed-bid uniform-price auction with a reserve price set at US$10 in 2012. The reserve price will rise by 5 per cent per annum above the prevailing inflation level.[43]

Offsets in the Californian cap-and-trade system must be permitted by CARB. Offsets can come from forestry, livestock and ozone-depleting substances. Entities may not use more than 8 per cent of offsets to fulfil their compliance obligation.

In addition to the use of offsets, the legislator has provided for the creation of a strategic reserve of allowances to contain high price increases. One per cent of the allowances from budget year 2013–14, 4 per cent from the second compliance period and 7 per cent from the third compliance period will be transferred to the 'allowance price containment reserve'.[44] The allowances from this strategic reserve will be auctioned if particular trigger prices are reached. These trigger prices lie, in 2013, at US$40, 45 and 50. These trigger prices increase by 5 per cent per annum plus the rate of inflation.[45] Furthermore, allocations to industrial entities are increased to account for their risk of carbon leakage.[46]

Every covered entity must comply with its annual and triennial compliance obligations by 1 November of the calendar year following the year for which the obligation is calculated and an emissions report has been submitted.[47] Non-compliance or the untimely surrender of allowances is subject to a penalty: for each missing allowance a total of four allowances must be surrendered.[48]

Banking is allowed but is subject to holding requirements that restrict the total amount of allowances that an entity may bank. Borrowing from future periods is not permitted.

---

43 Ibid., § 95911.
44 Ibid., § 95870(a).
45 Ibid., § 95913.
46 Ibid., § 95870 and Table 8-1.
47 Ibid., § 95856.
48 Ibid., § 95857.

California is a member of the Western Climate Initiative and has modelled its emissions trading system upon WCI design proposals. On 19 April 2013 the California Air Resources Board approved the linking of California's cap-and-trade system with that of Quebec. The linking should take effect from 1 January 2014.[49] It is conceivable that the Californian system might also be linked with the Australian emissions trading system.

Quebec

The Order of the Council 1187-2009 requires Quebec to reduce its emissions to 1990 levels by 2020.[50] The emissions for 2020 are set at 54.74 million tonnes of CO_2 equivalent. This implies an increased reduction of 3 to 4 per cent during the period 2015–20.[51] Emissions from all six Kyoto Protocol greenhouse gases (CO_2, CH_4, N_2O, HFCs, PFCs, SF_6) and NF_3 and other fluoridated greenhouse gases are covered.

The Quebec system started compliance operations on 1 January 2013 and is subdivided into different compliance periods. The first lasts until 31 December 2014. The second period will run from 2015 to the end of 2017, and the final three year period from 2018 to the end of 2020. At the end of each compliance period emitters must surrender emission allowances.[52]

In the first compliance period, around 80 sites from industrial and electricity sectors,[53] with annual GHG emissions equal to or greater than 25,000 metric tonnes of equivalent CO_2, are covered. With the beginning of the second compliance period on 1 January 2015, fuel distributers, and importers whose annual GHG emissions arising from its combustion reach or exceed the annual GHG emission equal to or greater than 25,000 tonnes of equivalent CO_2, will also be covered. As of 2015, the system will cover about 85 per cent of Quebec's total greenhouse gas emissions.

Emission reductions of greenhouse gases made during the period 1 January 2008 to 31 December 2011 can give rise to the free allocation of early reduction credits.[54] These early reduction credits are distributed for free and may be used for compliance purposes. Seventy-five per cent

---

49 California Air Resources Board (2013).

50 Order of the Council 1187-2009 of 19 November 2009.

51 Order of the Council 1185-2012 of 12 December 2012.

52 Environmental Quality Act, OC 1297-2011, s. 21; OC 1184-2012, s. 14, Art. 21.

53 Ibid., Annex A, Arts 2 and 19.

54 Ibid., Art. 65ff and Table B of Part I of Appendix C.

of the allowances estimated to be allocated free of charge via benchmarking formulae will be allocated to the respective emitter accounts on 1 May 2013 and 14 January of every compliance year.[55] The remaining 25 per cent of the total estimated quantity of allowances will be allocated for free on 14 September of every year after it has been determined whether an adjustment to the actual quantity must be undertaken.[56] Emitters not eligible for free allocation[57] must purchase their allowances at auction. The auction mechanism used for the Quebec system is a sealed-bid uniform-price auction with a reserve price set at CAD$10 in 2012. The reserve price will rise by 5 per cent per annum above the prevailing inflation level.[58]

Offsets in the Quebec cap-and-trade system may come from relevant manure storage facilities for methane (CH_4) destruction,[59] from landfill sites for CH_4 destruction,[60] and from the destruction of ozone-depleting substances contained in insulating foam removed from refrigeration and freezer appliances.[61] Entities may not use more than 8 per cent of offsets to fulfil their compliance obligation.[62] For every offset project, 3 per cent of the generated allowances are transferred to the Minister's environmental integrity account,[63] which can be used as a structural safeguard against illicit offsets.[64]

In addition to the use of offsets, the legislator has provided for the creation of a strategic reserve of allowances to contain high price increases.[65] This reserve can be used to supplement free allocation or sold by mutual agreement. The size of the reserve is set at 1 per cent of the emissions available under the cap for the years 2013 and 2014. For the period 2015–17 it is set at 4 per cent, and for the period 2018–20 at 7 per cent. Subsequent periods will have a reserve of 4 per cent. The sale by mutual agreement is open only to emitters established in Quebec and may be held four times a year if particular trigger prices are reached.[66]

---

55 Ibid., Art. 40.
56 Ibid., Art. 41 and Part II of Appendix C.
57 Ibid., Art. 39.
58 Ibid., Art. 49.
59 Ibid., Appendix D, Protocol 1.
60 Ibid., Appendix D, Protocol 2.
61 Ibid., Appendix D, Protocol 3.
62 Ibid., Art. 20.
63 Ibid., Art. 70.20.
64 Ibid., Art. 70.21.
65 Ibid., Art. 38.
66 Ibid., Arts 56 and 57.

The trigger prices for 2013 are set at CAD$40, 45 and 50, and will increase by 5 per cent annually above inflation levels.[67]

Every emitter must comply with its compliance obligations by 1 November and must have enough allowances in its compliance account to cover its verified emissions for every covered establishment during the years following the last compliance period for which emissions coverage was required.[68] Non-compliance or the untimely surrender of allowances is subject to an administrative sanction equal to three emission allowances per missing allowance. In addition to this penalty the missing allowance is retrieved also from the emitter's general account.[69]

Quebec is a member of the Western Climate Initiative and has modelled its emissions trading system upon WCI design proposals. Linking with other WCI member trading schemes will therefore not be problematic. The California Air Resources Board voted on 19 April 2013 to link its system with the Quebec system on 1 January 2014.

3.4 Midwestern Greenhouse Gas Reduction Accord

In November 2007, the states of Illinois, Iowa, Kansas, Michigan, Minnesota and Wisconsin and the Canadian province of Manitoba (now a member of the WCI) signed the Midwestern Greenhouse Gas Reduction Accord (MGGRA) to establish greenhouse gas reduction targets. The objective was to reduce GHG emissions by 60 to 80 per cent below 2007 levels[70] and to establish a cap-and-trade emissions trading system. The system was to be complemented by a GHG emissions reduction tracking system and other policies, such as low-carbon fuel standards. After presenting a model cap-and-trade system in 2010, the members of MGGRA stopped to pursue the objectives of their accord.

4. AUSTRALIA AND NEW ZEALAND

4.1 Australian Carbon Pricing Mechanism

After intense political debate, Australia joined the Kyoto Protocol and aspired to introduce a cap-and-trade scheme under the Kevin Rudd government. Rudd, however, failed to muster enough political support to

---

[67] Ibid., Art. 58.
[68] Ibid., Art. 21.
[69] Ibid., Art. 22.
[70] See Union of Concerned Scientists (2007).

pass the proposed scheme (known as the Carbon Pollution Reduction Scheme) and the bill was finally shelved in 2010. In February 2011, Prime Minister Julia Gillard's minority government successfully brought the Clean Energy Bill through both Houses of Parliament. This bill established the Australian Carbon Pricing Mechanism (CPM). This cap-and-trade scheme is an integral part of Australia's climate change policy and will help Australia to meet its mid-term and long-term objectives of reducing emissions by 5 per cent below 2000 levels in 2020 and up to 80 per cent below 2000 levels in 2050.

The CPM started on 1 July 2012 with a fixed carbon price of AU$23 per tonne of CO_2 equivalent emissions, which increases by 5 per cent per annum in real terms to account for inflation. During this fixed-price period companies purchase allowances – Australian Carbon Units (ACUs) – from the government for compliance. Energy-intensive trade-exposed (EITE) industries benefit from freely allocated emission allowances through the Jobs and Competitiveness Programme.[71] Since the Australian government sells as many allowances as demanded at the prevailing fixed price, the CPM does not have a cap on the number of allowances that are available. Banking and borrowing will be possible after 2014, although neither is allowed during the fixed-price period.

The CPM will transition to flexible prices on 1 July 2015. On the same day it will acquire an emissions cap and allowances will then be auctioned. Again, free allocation is available for EITE industries.

The Australian CPM was originally envisaged to operate as a price collar system. The price of emission allowances would have fluctuated within the bandwidth of a price floor and a price ceiling. The price floor was set at AU$15 and would have been linearly increasing by 4 per cent in real terms per annum. However, as a result of the linking negotiations between the Australian CPM and the EU ETS, the price floor was abolished.[72] The price ceiling still remains and will be set in 2014 at AU$20 above the expected EU ETS allowance price. It will increase by 5 per cent per annum.

The CPM covers approximately 300 of the largest emitters from stationary energy, industrial processes, fugitive emissions and non-legacy waste with direct greenhouse gas emissions of at least 25,000 CO_2 equivalents per year. The system covers the greenhouse gases of carbon

---

[71] In addition to this programme there is the Clean Technology Programme, which offers grants to EITE manufacturing industries to identify and implement technologies to improve energy efficiency and reduce their exposure to increasing energy prices. AU$1.2 billion is available under this scheme.

[72] See the Parliamentary Amendment Bill of 26 November 2012.

dioxide, methane, nitrous oxide and perfluorocarbons. It thereby covers around 60 per cent of Australia's GHG emissions. Non-covered sectors will be subject to an equivalent burden to incentivize GHG reductions. Emissions from agriculture and landholders are not directly covered but are eligible to take part in an offset scheme known as the Carbon Farming Initiative (CFI).

Within the framework of the CFI Australian Carbon Credit Units (ACCUs) can be earned for emission reductions or sequestration activities, and eligible ACCUs may be used for domestic compliance purposes. These ACCUs can be either Kyoto compliant (Kyoto ACCUs) or non-Kyoto compliant (non-Kyoto ACCUs). While Kyoto ACCUs are traded in both domestic and international voluntary and compliance carbon markets, non-Kyoto ACCUs are traded only in voluntary carbon markets.

In addition to ACCUs resulting from the CFI, other offsets may be used for compliance purposes in Australia. These are International Emissions Units (IEUs), deriving from the Kyoto Protocol's project-based mechanisms such as CERs, ERUs and RMUs.

During the fixed-price phase, up to 5 per cent of an entity's compliance obligations during a financial year may be made with eligible ACCUs. During the flexible-price period, no restrictions apply for ACCUs. International offsets can be used only in the flexible-price period, provided they are not temporary CERs, long-term CERS, ERUs or CERs from nuclear projects, large hydroelectric projects that are not in compliance with the World Commission on Dams guidelines, or stem from the destruction of nitrous oxide from adipic acid plants or trifluoromethane. The total amount of international emissions units is limited to 50 per cent of an entity's surrender obligations per compliance year. As a result of linking negotiations with the European Union, the proportion of European Union allowances (EUAs) permitted for compliance is set at 37.5 per cent, while the use of other international emissions units has been reduced to 12.5 per cent of an entity's compliance obligations per annum.

Entities falling within the CPM must surrender an eligible carbon unit for every tonne of CO_2 equivalent they emitted during a compliance year, which runs from 1 July until 30 June. To allow entities to compile their annual emissions reports, entities emitting more than 35 kilotons of CO_2 equivalent per year have two compliance dates. Every 15 June during the fixed-price phase entities must comply with what is known as the 'progressive' surrender obligation by submitting allowances equivalent to 75 per cent of their emissions during that year. The remaining 25 per cent of emission allowances are due every 1 February of the following compliance year. This is known as the 'true up' surrender. During the flexible-price phase there will be only one compliance date on 1 February following a compliance year.

Non-compliance with the surrendering obligations during the fixed-price period is subject to a penalty equivalent to 1.3 times the price of a carbon unit per tonne of CO_2 in a given compliance year. The penalty during the flexible-price period is twice the average price of carbon units in a given compliance year.

4.2 New Zealand Emissions Trading Scheme

New Zealand acceded to the Kyoto Protocol as an Annex I country and committed itself to stabilize its average emissions at 1990 levels by 2012. Conditionally it committed itself to reduce net emissions by 10 to 20 per cent below 1990 levels by 2020. By 2050 it aims to reduce its emissions by 50 per cent below 1990 levels.

The New Zealand Emissions Trading Scheme (NZ ETS) was introduced in 2008 and amended in 2009 to rally business support and to bring it in alignment with the Australian Carbon Pollution Reduction Scheme that was being proposed at that time.

The NZ ETS initially covered forestry sector emissions; it was extended in 2010 to stationary energy sources, liquid fossil fuels and industrial processes, and in 2013 to waste and synthetic greenhouse gases. While agriculture faces mandatory reporting obligations with regard to its biological emissions, its coverage is suspended indefinitely pending a review in 2015. With regard to the forestry and agriculture sector, the Ministry of Agriculture and Forestry is responsible for the management of the ETS; for all other sectors the Ministry for Environment is responsible.

The system is based on free allocation of New Zealand units (NZUs) in accordance with production benchmarks and current production levels. Provided that emitters surrender the corresponding amount of New Zealand units, or permissible emission allowances, industrial production levels can increase unhindered. The energy-intensive trade-exposed (EITE) industries and the agriculture sector receive different levels of free allocation. Agriculture will enjoy a baseline of 90 per cent free allocation and EITE industrial sectors will receive between 60 and 90 per cent. The level of free allocation is linearly reduced by 1.3 per cent every year. Entities that are not eligible for free allocation, mainly from the industry, have to buy emissions units from those that receive free allowances or from international markets or, of course, abate internally. This may change in the future as a new amendment authorizes the government to auction emissions units. Thus, in the future, the NZ ETS will have both auctioning and free allocation mechanisms.

The NZ ETS does not have an emissions cap nor any quantitative restrictions on the import of Kyoto credits. It allows the unlimited import of CERs, ERUs and RMUs. It does not allow temporary CERs, long-term CERs, ERUs or CERs from nuclear projects, the destruction of nitrous oxide from adipic acid plants or trifluoromethane. Since New Zealand is not a member of the second Kyoto compliance period, it will have more restricted access to Kyoto offsets in the future (that is, in 2015–16).

The NZ ETS has a one-year compliance period. Between 1 January and 31 March every year, covered entities must submit a self-assessment and audit report of their emissions over the preceding calendar year. Upon verification of the report, entities must surrender eligible emissions units by 30 April. Failure to do so adds a penalty of NZ$30 to the obligation to surrender allowances. The penalty may be waived if the defaulting entity informs the administering agency before the violation has been noticed.

The system has been subject to several changes. Some of these have already been implemented; others are yet to come. Transitional measures allow non-forestry operators covered by the NZ ETS to purchase NZUs at NZ$25 (the 'fixed-price-option') and to meet their surrender obligation by submitting only one tonne of CO_2 equivalent for every two tonnes they were obliged to surrender (the 'one-for-two' scheme). The transitional measures phase out progressively after 2012.

As an outcome of regular review of the scheme, a number of recommendations have been made. These recommendations include: (i) retaining the fixed-price option beyond the transitional phase until 2017, increasing its amount by NZ$5 annually and opening it up to new sectors that will join the scheme after 2012; (ii) extending the one-for-two surrender obligation beyond 2012 and progressively scaling it up to full surrender obligation by 2015; (iii) removing the ban on export of non-forestry NZUs once all covered entities assume full surrender obligations; (iv) a one-for-two surrender obligation for agriculture in 2015 and 2016 progressing to 67 per cent, 83 per cent and full surrender obligations in 2017–19 (if agriculture is covered); (v) revising the 1.3 per cent phase-out rate of free allocations to give an exact phase-out rate and the date on which free allocation ceases to exist; (vii) capping the emissions trading system.

The government has recently introduced the following changes to the system, some of which are based upon the review panel's recommendations.[73] The major proposed amendments include: (i) explicit powers on auctioning NZUs within an overall cap; (ii) phasing out the transitional measures more gradually; (iii) maintaining the fixed-price option until at least 2015 at the current level and providing for a power to extend it, if necessary, beyond that date and align it with the price ceiling of the Australian CPM; (iv) provide for the power to delay coverage of the agriculture sector indefinitely subject to a review in 2014–15.

5. ASIA

5.1 Japan

Following a period of increasing emissions in Japan, they contracted by 8 per cent in 2008 as a result of the economic downturn. They were only 1 per cent above 1990 levels at the beginning of the Kyoto compliance period. Under the Kyoto Protocol, Japan committed itself to a reduction of greenhouse gases by 6 per cent below 1990 levels during the period 2008–12.

Japan has been employing voluntary emissions trading schemes since 2005. These schemes did not have a fixed cap, were scheduled to finish in the fiscal year 2012, and did not yield major emission reductions at low cost.[74]

In 2006 Tokyo announced its long-term greenhouse gas reduction goals of reducing its emissions by 25 per cent below 2000 levels by 2020, and by 50 per cent by 2050. In June 2007 it announced the introduction of a cap-and-trade programme that would be conducive to achieving specific sectoral targets. According to the 2008 Environmental Master Plan, industry and commerce would reduce their emissions by 16 per cent below 1990 levels (17 per cent below 2000 levels), residential by 11 per cent (19 per cent below 2000 levels), and transport by 31 per cent (42 per cent below 2000 levels).[75]

The Tokyo Metropolitan Government Emissions Trading System (TMG ETS) was implemented in June 2008 and started operation in 2010. It focuses on CO_2 emissions from end use of energy in large buildings and factories covering around 1,300 facilities (commercial and

---

73 See New Zealand Government – Climate Change Information (2013).

74 Rudolph and Park (2010).

75 Tokyo Metropolitan Government (2010).

public buildings, and factories) above an annual consumption threshold of 1,500 kilolitres of crude oil equivalents.[76] CO_2 emission allowances are allocated for free based on fuel inputs and energy consumption, and verified by energy bills and external agencies. Process emissions as such are thus not covered. Emission rights are not allocated at the beginning of the compliance period but are used for calculating the amount of excess credits available to each facility.[77] As of 2011, only emission credits that are not needed for compliance may be traded.

The first two trading periods are from 2010 to 2014 and from 2015 to 2019, with an emissions reduction target of 6 per cent and 17 per cent respectively. During the first trading period factories and facilities using district heating and cooling systems are to be reduced by 6 per cent and commercial buildings by 8 per cent.[78] The base year for the free allocation is set as average emissions of three consecutive years between 2002 and 2007.

The TMG ETS takes early action of facilities into account.[79] 'Top level facilities', with outstanding energy-efficient equipment and superior performance, may be eligible for a reduction requirement that is reduced by 50 or 75 per cent. In addition, there is a new entrance reserve accounting for 0.74 per cent of the total allocation.

The TMG ETS allows for three types of offset, provided that these are verified by external agencies. Unlimited Emission Reduction Credits (ERCs) from energy savings by non-covered entities in the Tokyo area can be used. Renewable Energy Credits stemming from solar, wind, geothermal, hydro and biomass may also be used as offsets and are counted 1.5 times in order to encourage their use. Furthermore, ERCs from Japanese companies are accepted provided that they stem from large facilities with less than 150,000 of base year emissions.[80] The use of offsets from outside Tokyo is limited to one-third of the reduction obligation.[81]

The TMG ETS provides for a safety valve. If the market gets into turmoil the Governor of Tokyo is allowed to control the availability of

---

76 See Nishida and Jua (2011) p. 523.

77 Rudolph and Kawakatsu (2012) p. 10.

78 Nishida and Jua (2011) p. 523.

79 Early action was incentivized, inter alia, by the Tokyo 'CO_2 Emissions Reductions Program', which addressed large emitters: see Nishida and Jua (2011) p. 522.

80 See Rudolph and Kawakatsu (2012) p. 10.

81 Nishida and Jua (2011) p. 524.

offsets.[82] The Governor may increase the acceptance of domestic offsets, and also allow for the use of Kyoto Protocol offsets. This creates uncertainty regarding the price signal and the allocation of reduction measures.

Failure to comply with delivering emissions reduction or to submit reports is subject to a financial penalty of ¥500,000 and the obligation to cover the missing amount of emissions by a factor of 1.3. In the event of non-compliance the Governor will purchase the amount and be reimbursed by the facility. In addition, the name of the defaulting entity will be made public.

Banking of allowances from the first to the second trading period is allowed, while borrowing is not permissible.

Linkages between the TMG ETS and provincial schemes in Saitama, Kanagawa and Chiba are envisaged. The Saitama emissions trading system is, apart from its voluntary nature, modelled on the TMG ETS and it is therefore not surprising that Saitama allowances may be used for compliance within the TMG ETS. Given that the industrial structure in Kanagawa and Chiba is different, cooperation is more cumbersome.

5.2 China's Emissions Trading Systems[83]

China is the largest emitter of greenhouse gases in the world,[84] and has recognized climate change as an important challenge. In its 12th Five-Year Plan (FYP) (2011–15) China has set targets to reduce energy consumption per unit of GDP by 16 per cent, to raise the proportion of non-fossil fuels in the primary energy mix to 11.4 per cent and to reduce by 17 per cent the carbon intensity per unit of GDP below 2005 levels by 2015.[85] The fulfilment of these targets will allow China to make important progress towards honouring its commitment to reduce the carbon intensity per unit of GDP by 40 to 45 per cent compared with 2005 levels by 2020. The later commitment has also been included in the Copenhagen Declaration of the 16th Conference of the Parties[86] and is of particular relevance in the context of emissions trading.

---

[82] Ibid., p. 525.

[83] We are indebted to 刘冰玉, Master student at China EU School of Law, for very able research assistance.

[84] Cao (2011) p. 2.

[85] See Sandbag (2012) p. 12, and NDRC (2012).

[86] See NDRC (2010).

The 12th FYP is hailed as the beginning of a 'green era' in China.[87] Besides national measures for decarbonization, this Five-Year Plan has called also for the establishment of low-carbon zone pilot projects. Such zones have been established in five provinces (Guangdong, Liaoning, Hubei, Shanxi and Yunnan) and eight cities (Tianjin, Chongqing, Shenzhen, Xiamen, Nanchang, Guiyang, Baoding and Hangzhou).[88]

Independently of those low-carbon zones, the 12th FYP has mandated the establishment of seven emissions trading schemes in two provinces (Guangdong and Hubei) and five cities (Beijing, Tianjin, Shanghai, Chongqing and Shenzhen). These trading schemes have different GDP intensity targets.[89] In 2010 the cities and provinces accounted for around 19 per cent of China's population, 33 per cent of its GDP, 20 per cent of its energy use and 16 per cent of its CO_2 emissions.[90] After gaining experience in these zones, a national emissions trading system might emerge by 2016.

China does not only consider emissions trading as part of the instrumental mix to address climate change, but also a carbon tax. A few years ago, environmental taxation emerged as a hot topic in Chinese academic circles, with researchers suggesting the introduction of this form of taxation. Until now, the proposal still remains at research level. Environmental concerns have found their fiscal expression in China's draft Energy Law and its Renewable Energy Law. However, the provisions of those laws are criticized for not being sufficiently explicit to be implemented.[91] The carbon tax was first proposed in 2007 by the Institute of Fiscal Science, the Institute of Environmental Planning, the NDRC Energy Research Centre and Tsinghua University.[92] Interest in a carbon tax evolved only after considering environmental taxation and an energy tax. It is suggested that the recent interest and speed of deliberation of the carbon tax may, among other factors, be related to the animosities between China and the European Union regarding aviation.[93]

---

87 Sandbag (2012) p. 12.
88 NDRC (2010) p. 13.
89 See Sandbag (2012) p. 12.
90 See Jotzo (2013) p. 4.
91 We are indebted to Chen Ping, PhD candidate at the University of Ghent, Belgium.
92 Sandbag (2012) p. 14.
93 We are indebted to Chen Ping, PhD candidate at the University of Ghent, Belgium.

On 21 November 2011 a proposal for the introduction of a 10 RMB per tonne tax on carbon emissions was reported on Point Carbon.[94] Such a low level has been chosen in order not to disrupt economic growth, to allow for a gradual change of the business model and to enhance competition.[95] The tax is expected to include carbon-intensive businesses first before being extended. While this proposal may stem from a scientific report reflecting the government's opinion on a particular date, the actual text of the proposal could not be obtained. The actual validity and scope of the legislator's intent is thus still uncertain.

The sections below will present the information available on the seven Chinese emissions trading pilot schemes.

Guangdong Province

Guangdong Province is home to a high-energy consuming industry. Its pilot emissions trading scheme – as was announced in February 2013 – extends to an initial 239 trading companies and 310 reporting enterprises from four industry sectors, which include cement, steel, power and petrochemical. Guangdong Province will gradually expand the scheme at a later stage to more than 10 industries, including ceramics, textiles, non-ferrous metals, plastics, paper and other industry with annual carbon emissions of 20,000 tonnes.[96] The Guangdong implementation plan (released on 11 September 2012) is more detailed than that of other trading schemes in the country, and includes the following elements: (i) to establish the reporting and certification mechanism for carbon trading; (ii) to establish mechanisms for the management of carbon emission allowances; (iii) to establish the mechanism for the operation of carbon emissions trading; (iv) to carry out voluntary greenhouse gas emissions trading; (v) to explore the establishment of a mechanism for inter-provincial carbon emissions trading.[97] In addition, the Guangdong Development and Reform Commission has published rules on reporting to allow for the collection of emissions data.

The trading scheme is subdivided into several trading phases. The first runs from 2012 to 2015, the second from 2016 to 2020, with the third

---

[94] See Point Carbon (2011b).

[95] Sandbag (2012) p. 14.

[96] 'Implementation plan for Guangdong Province's carbon trading', available at: http://zwgk.gd.gov.cn/006939748/201209/t20120914_343489.html. For future information: http://www.nea.gov.cn/2013-02/26/c_132192742.htm.

[97] 'The first batch of allowances will be dispatched next month, Guangdong starts to establish carbon emission report system', available at: http://www.ccpit.org/Contents/Channel_50/2013/0226/356516/content_356516.htm.

trading phase commencing subsequently. Based on historical emissions during the period 2010–12 and taking into account the specific circumstances of the operator, Guangdong will issue the allowances for free, but this may be supplemented later by payment instruments. The first allocation took place in March 2013. In line with the macroeconomic position, and with reference to enterprises' reports of the previous year, Guangdong may make reasonable adjustments.

The decision to allocate allowances for free or to set a price will be taken by considering the province's total annual carbon emissions and abatement investment projects that have been undertaken exceeding 10,000 tonnes of standard coal.

Hubei Province

Hubei Province hosts a number of high-energy consuming industries. Its emissions trading system may therefore serve as a role model for provinces with a similar economic structure in the central and western regions of China.

Hubei Province will officially launch carbon emissions trading in August 2013. According to its implementation plan for the carbon emissions trading system, operators whose energy consumption exceeds 60,000 tonnes of standard coal equivalent per annum will be covered mandatorily.[98] There are more than 150 high-energy consuming enterprises from a vast array of sectors, including iron and steel, chemicals, cement, automobile manufacturing, electricity, non-ferrous metals, glass and paper-making. These enterprises accounted for 35 per cent of the province's carbon emissions and 52 per cent of the carbon emissions of the industrial enterprises in the province.

Given the scheduled launch of the Hubei emissions trading system in August 2013, a number of important steps are currently ongoing. In March 2013 attention was devoted to the formation of a carbon emissions trading centre, and in April on determining the actual emissions of operators; in May the focus will be on designing the emissions reporting and registration management system, as well as the system for electronic trading and settlements; in June attention will be on allocating allowances to operators.[99]

While, initially, only large emitters are included in the trading scheme, operators with an annual energy consumption of 8,000 tonnes of standard

---

[98] Jotzo (2013) (p. 47) reports in addition to the 60,000 tons of standard coal equivalent per annum also 120,000 tonnes of CO_2 per annum.

[99] 'Hubei Provinces will officially launch carbon trading in August 2013', available at: http://www.tanpaifang.com/tanjiaoyi/2013/0419/19604.html.

coal will also be monitored by independent accounting institutions and an annual carbon emissions report must be presented each year. These reports must be verified by a third party before submission. At a later stage, operators emitting less than 60,000 tonnes of standard coal may be covered by the emissions trading system.

If a covered enterprise fails to comply with its obligation within the prescribed period of time, it will face a triple penalty for the non-compliance margin and, at the same time, the allocated allowance will be deducted twice in the following year.[100] In the current phase, emissions trading will take place on the spot market only; a derivatives market may follow in the future.

There are several obstacles that impede the creation of an emissions trading system in Hubei Province. First, designing the appropriate target is difficult. Second, there is low enthusiasm on the part of enterprises. Third, there is incomplete data on operators' carbon emissions. Fourth, measurement and monitoring is still considered to be a challenge, and last, but by no means least, striking an optimal balance between environmental improvement and economic growth is difficult.[101]

Beijing City

The Beijing carbon emissions trading pilot will run from 2012 until 2015. The implementation plan has been submitted to the NDRC. According to this plan around 600 enterprises that emitted more than 10,000 tonnes of CO_2 per annum during the period 2009–11 are covered mandatorily. The allowances will be allocated on a yearly basis. In 2013, the emission allowance is based on the operator's emission levels during the period 2009–11. In 2014 and 2015 allocation is based on the emission levels of the previous year. Allocation of allowances will be largely for free, but the government has set a small part of the allowance aside for auctioning.[102]

Beijing's emissions trading system extends not only to direct CO_2 emissions stemming from the production of heat, electricity and combined heat supply for industries. It also covers indirect CO_2 emissions generated beyond the city borders for demand in Beijing: this includes

[100] 'Hubei steadily explores the way for Middle and Western regions to promote carbon trading', available at: http://info.china.alibaba.com/detail/1123025658.html.

[101] Ibid.

[102] 'Seven pilot provinces and cities explore carbon trading with Chinese characteristics', available at: http://www.sepacec.com/zhxx/xgxx/201210/t20121017_238852.htm.

some manufacturing plants and large public buildings. Beijing has established intermediary consulting, certification agencies and green financial institutions to support the trading market.

Tianjin City

The implementation plan for the Tianjin carbon trading pilot was released in February 2013.[103] In accordance with this plan, Tianjin will initially complete the construction of the emissions trading scheme in 2013. Its ambition for 2015 is to establish a regional emissions trading system and to become the national trading centre for emission allowances.

At present, Tianjin is focusing on establishing a verification system, on allocating allowances, and on establishing an allowance registration and supervision system. Enterprises emitting 20,000 tonnes of carbon dioxide per annum since 2009 will be included in the initial emissions trading pilot; so far, 120 large companies are covered, accounting for approximately 60 per cent of the city's CO_2 emissions.[104]

Allocation of allowances in Tianjin will take into account economic and industrial development, and the requirement of the carbon reduction target will be respected in determining the annual caps during the period 2013–15.

Annual allocation to each operator will be based on the annual emissions cap, the operator's past emissions and abatement measures that have been undertaken. Allocations may thus vary from year to year. In the future Tianjin will also introduce auctions to allocate allowances, and will also establish a compliance system. It will also study measures to impose constraints on those entities that do not comply with the law. However, specific measures on penalties are still lacking. Compared with those of Guangdong Province and Shanghai, Tianjin's implementation plan is more general.

Shanghai City

In November 2012 the first operators to fall within the Shanghai emissions trading system were identified: 197 operators from steel, petrochemicals, chemicals, non-ferrous metals, electric power and other industries with an annual CO_2 emissions rate exceeding 20,000 tonnes. In addition, sectors such as aviation, ports, hotels, financial institutions and other non-industrial companies with an annual CO_2 emissions rate

---

[103] 'Implementation plan of Tianjin's emission trading pilot', available at http://www.tjzfxxgk.gov.cn/tjep/ConInfoParticular.jsp?id=38237. For future information: http://www.chinatcx.com.cn/tcxweb.

[104] Jotzo (2013) p. 47.

exceeding 10,000 tonnes are covered. This might be explained by the structure of the economy of Shanghai: the city does not have a large energy-intensive industry. Around 600 enterprises are obligated to report their data.[105] Overall the system covers around 50 per cent of the city's emissions.

Allocation under the Shanghai scheme will take place once during the period 2013–15 free of charge. Free allocation is based on grandfathering and in cases where sufficient data exists on benchmarking: for the purpose of free allocation the context, expected growth and previous abatement efforts are taken into account.[106] Subsequently, an auction mechanism will be introduced.[107] The implementation plan, released on 3 July 2012, stipulates that allowances may not be borrowed during the pilot period, but they may be banked.

Shanghai will establish a carbon emissions monitoring, reporting and third-party verification system. Interestingly, Shanghai does not, so far, envisage fines for non-compliance in its plan. Further development of the details relating to management issues and regulations on emissions trading will take place during 2013–14.

It is worth noting that some finance institutions are also involved in the carbon trading market pilot: Shanghai Pudong Development Bank, for example, has officially announced that it will participate in the pilot. It has taken the initiative to plan domestic carbon financial and business services to assist the trading system.[108]

Chongqing City

Chongqing's implementation plan for carbon emissions trading has been approved early 2013. This pilot scheme is concentrated mainly in the electrolytic aluminium, ferro-alloy, calcium carbide, caustic soda, cement, steel and other high-energy consuming industries. The carbon trading market is expected to be operational in September 2013, and is

---

[105] 'Implementation plan of Shanghai's carbon trading', available at: http://www.shanghai.gov.cn/shanghai/node2314/node2319/node12344/u26ai32789.html. For future information, visit http://www.cneeex.com.

[106] Jotzo (2013) pp. 47–8.

[107] 'Shanghai launches carbon trading pilot and the initial allowances are charged for free', available at: http://sh.eastday.com/m/20120817/u1a791296.html.

[108] 'Shanghai Pudong Development Bank has determined to participate in the carbon trading', available at: http://bank.hexun.com/2012-11-26/148360941.html.

then expected to expand gradually in scope and scale of transaction.[109] The covering threshold is set at 20,000 tonnes of CO_2 per annum for industrial installations and at 10,000 tonnes of CO_2 emissions per annum for non-industrial sectors.[110]

Shenzhen City

In September 2012 the first 117 enterprises to take part in the Shenzhen emissions trading system were announced. In November 2012 an additional 300 enterprises were identified and are undergoing the verification process. A number of operators in the first cohort have already been through the verification process and have provided information for the next step of data analysis. Thresholds for inclusion were planned to be set at 20,000 tonnes of CO_2 per annum but this figure will be revised downwards to 10,000 tonnes.[111] Compared with the other pilot cities, Shenzhen's emissions trading system does not cover steel, cement and other large emission sources. The city has already established mandatory emission reduction targets for the 800 key energy consuming enterprises. After several rounds of verification, 635 enterprises are included in the carbon trading system, which accounts for 40 per cent of Shenzhen's total carbon emissions in 2010.[112] With this, the city has the largest number of enterprises participating in the scheme, compared with other pilot emissions trading systems in China. Shenzhen will officially embark on the trading market in June 2013.[113]

6. CONCLUSIONS

This chapter has presented several emissions trading systems that operate in different parts of the world. The object and purpose of this review was not to deliver a complete and detailed description of each system, but to

---

109 'Chongqing will launch carbon trading, high energy consumption industries are involved', available at: http://www.coal.com.cn/Gratis/2012-11-21/ArticleDisplay_329613.shtml. For future information see http://www.cquae.com/html/tpfjy.

110 Jotzo (2013) p. 47.

111 Ibid., p. 47.

112 'The pilot carbon trading is undergoing the last stage, how to design allowances becomes the focus', available at: http://szsb.sznews.com/html/2013-04/04/content_2431510.htm.

113 'Shenzhen carbon trading will be launched in June', available at: http://news.xinhuanet.com/politics/2013-04/01/c_124528467.htm. For more information: http://www.cerx.cn/cn/index.aspx.

present their diversity. It is evident that the various jurisdictions have been finding very different solutions for common challenges. From a design perspective some interesting observations can be made.

1. It is noticeable that there is some form of pressure to conform on some of the emissions trading schemes. The Swiss and the Australian schemes have been adapting their design features along the lines of the EU ETS. This is not surprising in light of the ongoing linking negotiations between these countries and the European Union. What is important to note also is that New Zealand has altered its system on several occasions, most likely as a response to the changes in the Australian system. New changes are currently on the way in New Zealand and it may be wondered if its ETS will move eventually towards the European design. In this sense EU climate leadership would indeed generate followers – perhaps not entirely voluntarily – but this does not discredit the proliferation of a particular ETS design.
2. There is a remarkable difference among environmental targets across the various systems. This will make a comparison of climate change efforts difficult. China (not included in Table 4.1 below since it has too many emerging systems where too much information is yet to be revealed) sticks out since it has adopted an intensity target and thereby uses a very different approach from most systems.
3. An element that will make comparison of the cost burden on industry and its competitiveness difficult to assess is that the thresholds for coverage differ considerably.
4. The types of emission included in the various systems differ. Some systems focus only on CO_2 emissions, other systems on the six greenhouse gases, and yet others include additional gases.
5. It can be observed that a large variety of allocation mechanisms are used, including free allocation, benchmarking and auctioning.

However, among the different emissions trading systems there are similarities. It can be pointed out, for example, that most schemes find means to support their industry, and will impose sanctions to coerce compliance.

Table 4.1 below highlights some of the elements discussed in this chapter.

Table 4.1 Comparison of some existing emissions trading schemes

	EU ETS	Swiss ETS	RGGI	California ETS	Quebec ETS	Australia CPM	New Zealand ETS	Tokyo ETS
Aim	2020: 21% below 2005 levels	2020: 20% below 1990 levels	2018: 10% below 2009 levels	2020: -3% to -4% during 2015–20	2020: -3% to -4% during 2015–20	2020: 5% below 2000 levels	2020: 10% to 20% below 1990 levels	2020: 25% below 2000 levels
Gases	Multiple GHGs	Multiple GHGs	CO_2	Multiple GHGs	Multiple GHGs	Multiple GHGs	Multiple GHGs?	CO_2
Threshold	20 MW	20 MW	25 MW	25,000 metric tons	25,000 metric tons	25,000 tons		1,500 kilo litres of crude oil
Sectors	Energy + industry	Energy + industry	Energy	Energy + industry	Energy + industry	Energy + industry	Energy + industry (forestry in the future?)	End users of large energy buildings, commercial and public buildings, factories
Allocation	Benchmark + auction	Benchmark + auction	Auction	Benchmark + auction	Benchmark + auction	Free allocation + auction	Free allocation + buy-up	Free allocation
Carbon leakage	Compensate indirect emission costs	Up to 100%	Cost containment reserve	Allowance price containment reserve and increases in allocation	Strategic allowance reserve	Free allocation for EITE companies, price ceiling	Free allocation up to 90%	Offset use increase
Offsets	Limited import	Limited import	Limited import	Limited import	Limited import	Limited import	Unlimited import (reduced access to Kyoto offsets)	Limited use
Sanctions	€100 + reparation	CHF125 + reparation	3 rights for 1 ton	4 rights for 1 ton shortfall	3 rights for 1 ton + reparation	1.3 times allowance price	NZ$30	¥500,000 + 1.3 times reparation

Source: Author's own representation.

5. Implementation issue 1: Initial allocation of emission rights[1]

1. INTRODUCTION

Before starting emissions trading, an emissions target has to be set. This can be formulated in terms of an absolute emissions cap or in terms of a relative emissions target (expressed, for example, per tonne of CO_2 per unit of output or per unit of GDP). In order to reduce emissions, this cap or relative target must be more stringent than current emission levels, otherwise the regulation will not 'bite'. The next step is to determine the way in which emission rights will be allocated. Emitters either obtain these rights for free or they have to buy them from the government (for example, at auction). ETS designers could also opt for a combination of measures to bring allowances onto the market.

Within these categories of free or paid allocation various options can be used to bring allowances onto the market. Standard forms of free allocation employed by emissions trading systems are grandfathering and benchmarking. Grandfathering refers to the allocation of emission allowances on the basis of historical emissions. Benchmarking refers to an allocation based on production technology standards. As a standard form of paid allocation, auctions are commonly used. Other forms of paid allocation (such as the sale of allowances to a specific section of society) are theoretically possible, but are not employed in practice in the context of the initial allocation of allowances. They are, however, employed to bring additional allowances onto the market when emission allowance prices are too high: the Australia Carbon Pricing Mechanism, for example, contains a price ceiling system. The usual form of 'selling' allowances onto a market is by means of allowance auctioning, and auctions can take various forms. Each of these allocation options has different consequences in terms of efficiency, effectiveness and acceptance. Each initial allocation has its

[1] This chapter was written mainly by Edwin Woerdman, who has kindly permitted its use in this book.

specific winners and losers, which can give rise to intense and costly (and therefore welfare reducing) lobbying.

This chapter is organized as follows. Section 2 discusses target setting in an ETS, including the potential problem of over-allocation as has been experienced in the European Union ETS. Section 3 explains free allocation based on either grandfathering or benchmarking under an emissions cap, as well as free allocation in the absence of an emissions cap. It also considers the 'windfall profits' discussion, which was prominent in the EU ETS for some time and necessitated the change of allocation mechanisms. In Section 4 different auction design variants are analysed. Section 5 presents a number of examples of lobbying, as well as its economic consequences, focusing again on the EU ETS. A conclusion is presented in Section 6.

2. SETTING THE EMISSIONS CAP

Emissions trading is based either on an emissions cap, or on the setting of relative targets which have to be formulated and adopted before trading in emission rights can begin.

A cap-and-trade system imposes a cap on the annual emissions of a group of entities for a period of years. This total emissions cap is divided into smaller units of allowed emissions per installation by giving away emission allowances for free or by selling those allowances. All allowances are tradable. At the end of the compliance period (one or more years) each operator must surrender a number of allowances equal to its emissions during that period. Non-compliance results in a costly penalty.

Relative targets can be enforced via credit-and-trade schemes. Here mandatory emissions standards are set for a group of companies. The emissions standard dictates permitted emissions per unit of energy consumption or per unit of GDP. Tradable credits can be earned by emitting less than the amount prescribed by the emissions standard, and can be sold to those operators that are unable to comply with the emission standard. Penalties will be imposed on those operators that do not comply with the production standard and fail to surrender the required amount of emission credits.

2.1 Over-allocation

The emissions cap or the emission standard of a relative target scheme must be more stringent than 'business as usual' emission levels. If the policy targets are too lax, the required scarcity of emission allowances is not created; there will therefore be no positive market price for such allowances, and hence no incentive to abate emissions.

This happened, for instance, in the first trading phase of the EU ETS from 2005 to 2007. The average allocation ranged from about 105 per cent of actual emissions in the cement sector to about 120 per cent in the pulp and paper sector.[2] In early 2006 it became clear that the number of emission allowances allocated in the EU as a whole was 4 per cent above actual emission levels.[3] 'Over-allocation' became the term used in the literature to label this situation. A number of articles sought to determine which countries were allocating more than needed and how sectors were affected.[4] Woerdman et al. classify the EU Member States into three categories:[5]

- countries that over-allocated allowances to their national industries, which undermines the incentive to trade in emissions – namely Austria, Denmark, Finland, France, Germany, Greece, Ireland, Italy, Luxembourg, the Netherlands, Portugal, Spain and Sweden;
- countries for which the position is not entirely clear – namely Belgium and Slovenia;
- countries that under-allocate allowances to their national industries, which is a correct prerequisite for emissions trading – namely the Czech Republic, Estonia, Hungary, Lithuania, Latvia, Slovakia and the United Kingdom.

One of the causes of over-allocation in the EU is attributable to policy weaknesses that are often referred to in law and economics literature as 'government failure'.[6] Over-allocation was the result of incomplete

---

[2] Trotignon and Delbosc (2008).

[3] Ellerman and Buchner (2006).

[4] Clò (2007); Kettner, Köppl, Schleicher and Thenius (2007).

[5] Woerdman, Clò and Arcuri (2008).

[6] Law and economics literature distinguishes between government failure and market failure. Market failure refers to instances where the market fails to provide an item or service at all, or only at higher costs than would have been the case had the market been functioning efficiently. Examples of market failure are imperfect competition, external effects, collective goods and information asymmetries. The government then needs to step in with regulation, for example, to improve competition in the market, to internalize externalities, to provide collective goods and to reduce information assymetries. Government failure then refers to situations where this regulation is imperfect and leads to inefficient outcomes. Examples of government failure are policies and measures that lead to

information about actual historic emission levels of operators and over-optimistic predictions of production growth.[7] The occurrence (and magnitude) of such an 'emissions gap' resulting from lower than predicted emissions of economic activity was unexpected. That it was unexpected not only for governments, but also for operators, can be inferred from the steep increases in emission allowance prices at the beginning of the trading period.

However, another cause of over-allocation in the EU is 'government failure' as a result of industry lobbying. For lobbying to work 'it takes two to tango': not only lobbyists that try to influence environmental policy, but also a government that is willing to implement (some of) their wishes. The problem was that every operator wanted to be a seller rather than a buyer of allowances. Each industry sector thus had an incentive to lobby for as many allowances as possible. A generous allocation of allowances increases the acceptability of emissions trading at the level of individual firms, but the collective result is an inefficient scheme in which the allowance price will be too low compared with the costs of climate damage resulting from industry emissions.

In theory, if there is an over-allocation of allowances (and if banking is not possible), the allowance price could even drop to virtually zero. In practice, this is what happened with the EU ETS: the allowance price dropped from €30 to around €10, and then went down to €0.10 in 2007.[8] The excessive supply of allowances implied that an emissions trading market had been created in the EU without scarcity. If a net over-allocation of allowances occurs, emissions are capped but the emissions trading scheme will not provide incentives for emission reductions and will lead to political debates over the scheme's environmental effectiveness.

Some newspapers stated that the EU carbon market was failing because of the spectacular price drop, but the opposite is true. The carbon price response to over-allocation indicated that the emission allowance market was functioning properly. If there is no shortage of emission

---

incorrect incentives, overproduction, overconsumption, operational slack, lack of choice and lack of innovation. These concepts will be used in this chapter.

[7] Although it should be noted that the European economy grew by 2.1% in 2005, by 3.3% in 2006 and 3.2% in 2007. See EU GDP growth rate in real terms, values for EU 27, available at: http://epp.eurostat.ec.europa.eu/tgm/table.do?tab=table&init=1&plugin=1&language=en&pcode=tec00115 (last accessed 27 May 2013).

[8] The price of €0.10 probably reflects the transaction costs incurred in selling allowances.

allowances, the price will be virtually zero. The carbon price drop as a result of over-allocation was not caused by 'market failure' but by 'government failure'.

A problem related to this over-allocation of emission allowances to covered sectors will arise if the overall greenhouse gas emissions of a country or jurisdiction are 'capped'. Some countries have assumed such national emissions targets under the Kyoto Protocol. In the presence of a binding national emissions cap, failure to set adequate emission reduction targets for the emissions trading scheme will constitute a benefit for the ETS sectors. The environmental costs are shifted from the ETS sectors to the non-ETS sectors. The reason for this is that sectors outside the scheme will have to increase their reduction of emissions in order to enable the country to comply with its national emissions target. If the abatement costs of non-covered sectors exceed the abatement costs of covered entities, such subsidization is inefficient. In addition to efficiency concerns, it may also be perceived as 'unfair'.

Equalizing marginal abatement costs between sectors covered by an ETS and sectors that are not covered but are subject to other policy instruments is inherently difficult. In the EU, for instance, marginal abatement costs are higher for non-ETS sectors such as transport, households and agriculture.[9] The implication is that the ETS sectors should bear a higher (and not a lower) emission reduction burden than the non-ETS sectors.

The government could compensate for over-allocation in the emissions trading system by acquiring credits from Joint Implementation (JI) and Clean Development Mechanism (CDM) projects. Taxpayers would be required to pay for those emission allowances. This could transform the international Kyoto mechanisms into largely publicly funded markets.[10] In this situation, over-allocation constitutes a subsidy to the ETS sectors since governments would then abate on behalf of the sectors covered by the emissions trading system in order to comply with their national emission reduction targets.

With over-allocation, the environmental costs are only partly internalized by polluters in the emissions trading scheme as the remaining costs are shifted (or 'externalized') to those outside the scheme. This means that over-allocation violates the polluter-pays principle.[11]

---

[9] Criqui and Kitous (2003); Böhringer et al. (2005); Peterson (2006).

[10] Hepburn et al. (2006).

[11] Woerdman, Clò and Arcuri (2008).

In sum, over-allocation is not only ineffective, but is also likely to be inefficient and unfair. The environmental costs that the ETS sectors should bear to contribute towards achieving the national emissions target are shifted to non-ETS sectors, which are subject to other policy instruments and for which abatement costs are usually higher. The harmful effects of over-allocation on the efficiency, effectiveness and fairness of an emissions trading scheme could undermine the scheme's political acceptability.

2.2 Options to Tackle Over-allocation

There are several options that emission designers could employ to counter over-allocation of initial allowances, which will be addressed here. We will also briefly examine ways of addressing situations of over-supply. First, measures that could be considered to address primary over-allocation will be discussed; those are situations where allowance supply is so large that the allowance price is virtually zero. Second, we look at measures that can be employed in situations where there is oversupply, but still a positive market price; those are situations where there is too great a supply of allowances on the market to deliver a sufficiently high price to incentivize technological innovation. Such situations may stem from an economic downturn, an influx of offsets to the trading scheme or from supply shocks in one of the covered industry sectors (such as the price decline of gas as a result of the availability of shale gas). All of these effects will lead to a reduction in the emission allowance price.

Where it is apparent that over-allocation has been occurring in the primary allocation of allowances, the market price will tend to drop to virtually zero. The government can start to buy up allowances until the price recovers. This is a direct support scheme for the industry, and at times of fiscal austerity is unlikely to be popular with the electorate. From an environmental effectiveness perspective, however, it would be a quick remedy for the problem, and would allow for the effective functioning of the allowance market and the realization of abatement. If such policy measures are not feasible, then a 'quantity'-based solution must be found.

Another effective way to address over-allocation would be to stop the compliance period as soon as possible and then re-allocate allowances based upon documented emissions data. Such drastic measures give rise to significant administrative costs and, from a legal viewpoint, are unlikely to be capable of implementation since the required legal amendments would not be in place in time.

There are, however, other policy options. Allowances can be deleted or debased. Debasing the face value of an allowance means that its value could be reduced to less than one tonne of greenhouse gas emissions. The underlying idea is that, just as money is subject to inflation, so too allowances could become worth less. Deleting allowances or debasing them would certainly be unpopular with covered entities since it would be tantamount to an economic expropriation. Whether such policy choices would be permissible from a legal expectation perspective, and whether they would be compatible with the legal system, would need to be assessed. Depending on the 'property right' definition of an emission allowance, these may not be viable legal options.

Since it is difficult to remedy a situation of initial oversupply, emissions trading designers should seek to avoid such situations. In the event that the necessary data for setting up an emissions trading system is unavailable, designers should consider allocating fewer allowances for free and making additional allocation available if the allowance market price reaches predetermined trigger prices.

In contrast to situations of primary over-allocation, situations of oversupply are more easily addressed. In its second trading phase, the EU ETS is suffering from a significant amount of oversupply that is derived from the economic downturn in 2009 and from an increased use of offsets. The market price of EU emission allowances is still positive (at €3.41 in May 2013),[12] but too low to incentivize investment in carbon-friendly technologies. So, in the second trading phase, the EU ETS is not suffering from over-allocation but rather from excess supply. It is, however, an approach that could also be applied in the context of a situation of over-allocation.

Since banking is permitted between the second and third trading periods of the EU ETS, the oversupply situation is projected to continue well beyond the third phase. The European Commission has proposed postponing the auctioning of allowances towards the end of the trading period (2019 and 2020). This 'backloading' of allowances does not correct the market, but it allows the European Commission to introduce new legislation to eventually delete the 'backloaded' allowances. Depending on the design of the policies, they will not constitute a systemic solution to the problem of economic downturns, but will only represent an ad hoc solution which neglects the conventional wisdom that economies move in business cycles and that economic downturns are a recurring issue.

---

[12] EUA spot market price on 27 May 2013: see http:/www.eex.com.

An alternative to reducing the supply of emission allowances in cases of limited over-supply is to permit the use of offsets once a particular trigger price is surpassed. In this way a market that is already over-supplied will, because of the low prices, not be flooded with additional allowances. The Australian CPM used such a system. Albeit effective to support the price on the domestic allowance market, it is inefficient from an economic perspective since it prevents the exploitation of cheap abatement opportunities abroad.

Reserve price auctions, combined with a mandatory cancelling of allowances that are not being auctioned, will not hinder the access of domestic enterprises to cheap abatement opportunities that are available outside the system. It would therefore constitute a better way to address oversupply situations – provided, of course, that the amount of the allowances auctioned in the domestic market is sizable.

3. FREE ALLOCATION OF EMISSION RIGHTS

3.1 Free Allocation under an Emissions Cap

Free allocation of allowances under an emissions cap can be based on either:

- grandfathering, or
- benchmarking.

Grandfathering refers to free allocation of allowances to enterprises based on their historical emissions, whereas *benchmarking* refers to free allocation of allowances to enterprises based on 'clean' production standards. Basically, grandfathering gives emission rights to 'dirty' firms; with benchmarking emission rights are given to 'clean(er)' firms. This has no impact on the environmental effectiveness of the emissions trading scheme, but it constitutes a distributional choice that can be made to enhance political acceptance.

Grandfathering should be chosen if vested interests of incumbent firms that have not yet invested in clean production technology are to be protected. Grandfathering on the basis of historical emissions takes due account of the magnitude of their emissions, and will allow them to continue to emit more in the future, although it should be observed that the actual pollution and abatement decision will not be based on allocation but on the prevailing market price of emission allowances.

If, by contrast, incumbent enterprises have invested in clean production technology and they stand to compete with those that use inferior technology, their interests would best be served by benchmarking. In the context of the EU ETS, the Dutch chemical industry, for example, lobbied strongly for the adoption of an allocation format that would grant it an advantage vis-à-vis its regulated competitors in other EU Member States.

In the EU ETS, politicians' perception of what was a 'fair' allocation evolved over time. It moved from grandfathering to benchmarking. It should be noted that grandfathering is also capable to taking into account early investments by incumbent firms, referred to as 'early action'. This was done in the EU ETS through allocation formats whereby allowances were allocated based on historical emissions, but a correction factor was introduced by various Member States to reward firms with low emissions per unit of energy or output.[13] Alternatively, allowances can be allocated based on product benchmarks, where a correction factor can be used to reduce the allocation for products with no risk of carbon leakage.

Whatever allocation format is used, in one way or the other emitters need to receive a limited number of emission rights, and greenhouse gases need to be priced by the market. In practice, however, benchmarking may lead to a tighter allocation of allowances. Lecourt et al.[14] estimate that an additional 670 million allowances will be available to be auctioned by public authorities under the EU ETS compared with a hypothetical continuation of the earlier grandfathering rules. To a large extent, this can be explained by a reduction in the over-allocation of allowances. A larger share of allowances to be auctioned enhances the efficiency of the EU ETS, since auctioning is in some respects more efficient than free allocation, as will be explained below in this chapter.

3.2 Free Allocation under a Credit-and-Trade System

A credit-and-trade system operates in the absence of an emissions cap. Policy targets are set in relative terms and emission credits are created by over-complying with a set emissions standard. Enterprises acquire these credits for free. Those that fail to comply with the standard must purchase emission credits and surrender emissions to the government in accordance with their actual emissions.

---

13 See, for example, NAP-NL (2007).

14 Lecourt, Pallière and Sartor (2013).

In the literature such a credit-and-trade system is often referred to as a 'performance standard rate' (PSR) system. Under a PSR trading system, emission credits are available for free and are arguably not being allocated at all. The reason is that credits come into existence 'bottom-up' only by reference to a production standard without the actual 'top-down' allocation of emission allowances. This distinction can be useful in the context of legal disputes under European or international law.[15]

Credits are always made available for free. Auctioning these credits, although theoretically possible, would run counter to the economic efficiency as well as the environmental objectives of the PSR system, and hence go against the very motivation for adopting such a trading system in the first place.[16]

Under a PSR system, an emitter is allowed to emit as long as it complies with the emissions standard. If the government has adopted an intended emissions target, it can sharpen the emissions standard in order to keep the overall amount of emissions within this target. If the economy grows, for instance, the number of emissions also increases because companies do not have an emissions ceiling but have to comply with the emissions standard. If an energy-intensive entity wishes to expand production, or if a newcomer enters the industry, each has a right to emit. In contrast with a cap-and-trade scheme, an enterprise operating under a PSR system does not need to purchase allowances from existing emitters or from a government reserve in order to be able to emit. Instead, it acquires its emission credits by over-performing with the set production standard.

The first implication of this is that total emissions can rise, even if the emissions standard itself is complied with. This could lead to emissions beyond the national or sectoral target if such a goal has been formulated. The second implication is that the social costs of the emissions exceeding a formulated environmental target are not fully reflected in the costs per unit of production. If the emissions standard is not automatically or imperfectly adjusted, this particular emissions trading design could therefore be less environmentally effective than emissions trading under an emissions cap.

However, a PSR system may be chosen to accommodate political concerns. Free allocation under an emissions standard has advantages for certain interest groups, such as firms which do not have to purchase extra

---

15 Weishaar (2007b).

16 Weishaar (2012) p. 100.

emission rights if they seek to expand their production. Moreover, under a cap-and-trade scheme an additional choice must be made between auctioning allowances or giving them away for free.

3.3 Windfall Profits

Basic microeconomics teaches us that the legal incident of collecting taxes is logically separable from the economic incident of who actually has to bear the cost, which falls predominantly on those parties who can least avoid it.[17] Who eventually bears the cost burden is an issue of 'bargaining power between parties' (economists call this 'demand elasticities') and is consequently independent of the type of cost.[18]

Politically, however, free allocation of emission rights led to a political debate about 'windfall profits', as witnessed in Europe at the start of the EU ETS in 2005. In order to make emissions trading acceptable for firms, a political choice was made not to auction the allowances but instead to grandfather them (free allocation, based on historical emissions). An unexpected consequence of this, however, was the political turmoil that arose in the first year of the scheme's operation over electricity companies passing on the carbon price to large electricity consumers (such as energy-intensive industries) as well as to small electricity consumers (such as households). Although electricity companies receive their allowances for free, they added the carbon costs to the consumers' electricity bills. These companies were then accused of making additional profits, referred to as 'windfall profits'.

Jepma, for instance, has written:

> Because allowances are grandfathered, and most installations seem to succeed in passing on allowance prices onto end users without being charged accordingly, they eventually capture windfall profits. ... [T]he group that eventually pays for the rent ... , is the group of final end users, or consumers, who eventually pay the bill for the net windfall gain.[19]

From an economic perspective, however, there is nothing wrong with this.

In fact, economic theory teaches that the electricity companies were behaving correctly by passing on the carbon price to consumers.[20]

---

17 Frank (1997) p. 52ff.
18 On this issue see Weishaar (2009) p. 182.
19 Jepma (2006) pp. 6–7.
20 On this issue see Woerdman, Couwenberg and Nentjes (2009).

Windfall profits are not profits; they are costs. To provide a more general background to this reasoning, it is important to understand that, in economics, the concept of 'opportunity cost' must be taken into account whenever a resource can be used in alternative ways.[21] When you consume more of item 1, you may have to give up some consumption of item 2. Giving up the opportunity to consume item 2 is the economic cost of more consumption of item 1. An example is the wage rate: this is not simply the price of labour; it is also the opportunity cost of leisure. If your salary is €20 per hour, then an extra hour of leisure costs you €20 in forgone income.

Free allocation thus entails costs for companies, namely the opportunity costs of the allowances when they are used for covering emissions. Instead of using the allowances, the company could have sold those emission rights. Independently of the initial allocation, profit-maximizing firms will therefore take into account the opportunity costs[22] of the awarded emission allowances and rationally decide to use or to sell them. This is not a 'market failure', as some critics have stated, but reflects upon the orderly working of the market. Profit-maximizing enterprises will pass on the opportunity costs to the best of their abilities. Because electricity producers are in a position to pass on a large part of the production costs, consumers have to pay for the allowances that energy producers obtained for free and the energy producers obtain windfall 'profits'.

Before the start of the EU ETS, European politicians decided to use free allocation of allowances in order to make carbon pricing acceptable for their industries and to protect their countries' competitiveness. Interestingly, once the scheme was up and running, politicians were surprised to see that electricity companies made windfall profits. This should not have come as a surprise at all. The free allocation of emission rights generates assets on the balance sheet and thereby improves the financial position of shareholders. The value of a share increases because the electricity company receives an asset with a market value for free. What is really surprising is that the politicians did not anticipate the incentives of firms to pass on such costs to their consumers.

---

[21] See, e.g., Varian (2003).

[22] Opportunity costs are commonly understood as the cost of something measured in terms of the benefit forgone by not using it for a different purpose. For a straightforward introduction to the relationship between opportunity costs and free allocation of emission allowances see Woerdman, Couwenberg and Nentjes (2009).

However, the increase in the cost price of electricity by fully incorporating the opportunity costs of free allowances is only partly reflected in a higher market price for electricity. In the EU, Member States such as the Netherlands, Germany, the United Kingdom and Finland, pass on rates varying roughly between 30 per cent and 100 per cent.[23] The reason for this phenomenon is the oligopolistic structure of the electricity market. Some politicians in Europe thought that windfall profits appeared because of anti-competitive behaviour by power companies, and that those profits would go away by making the electricity market more competitive. Two arguments presented in the literature claim that this opinion is economically unsound.[24] First, the passing on of the opportunity costs of free allowances to consumers is not a consequence of too little competition in the electricity market, but a consequence of free allocation in the carbon market. Second, making the electricity market more competitive might increase windfall profits, since prices would then be more aligned with (opportunity) costs.

3.4 Options for Tackling Windfall Profits

A year or two after the start of the EU ETS, several ideas for addressing windfall profits were discussed.[25] The policy options considered were: (i) strengthening the emissions caps; (ii) forbidding the mark-up; (iii) taxing the windfall profits; and (iv) auctioning the allowances. These four options will be discussed briefly below.

The first political response to the electricity companies' windfall profits by a number of EU Member States (for instance, the Netherlands) was to allocate relatively less to the power sector for the period 2008–12. In addition, the emissions cap was tightened (but, as a result of the economic downturn, the market turned out to be oversupplied). The price of emissions rights should therefore have been higher than they actually were during the second trading period. As a consequence, windfall profits for the power sector would have been reduced but would still have persisted. To alleviate this problem, these EU Member States (including the Netherlands) distributed relatively more emission rights to the other covered sectors in the scheme. The Dutch government, for example, allocated 15 per cent fewer rights over the period 2008–12 to the electricity sector, gave one-third of those unallocated rights to non-electricity participants and sold two-thirds on the carbon market. The

---

23 Gullì (2008).

24 Sijm et al. (2005) pp. 98–103.

25 Woerdman, Couwenberg and Nentjes (2009).

proceeds of those sales went to energy consumers. Although this redistribution policy may sound well thought-out, windfall profits did not disappear by strengthening the emissions caps and redistributing allowances. This is no surprise if one recognizes that free allocation and elasticities are the cause of windfall profits.

A second option is for politicians to introduce price regulation in order to prevent price increases. If the market value of free allowances is not allowed to translate into price increases in electricity bills, the producer might hypothetically be better off selling both its power station and its allowances than to continue to produce electricity. However, the incentive to exit the market is reduced if the emissions trading scheme also has a rule that a producer loses its allowances when an installation is closed down (closure rules are discussed later in this chapter). Apart from high administrative and information costs to the government, regulating prices, in particular in the context of imperfect information available to governments, will be inefficient and will reduce social welfare.[26]

A third idea is that politicians could tax electricity producers for the windfall profits they make. Such a windfall profit tax is administratively problematic because the opportunity costs of free allowances cannot be found in consumers' electricity bills, nor are they listed as such in the company annual accounts. Producers have an incentive to hide such windfall profits as costs in their accounts, or to adjust asset values in calculating the accounting (not windfall) profits. Moreover, if the mark-up on the electricity price is taxed away, electricity producers receive no compensation for the opportunity costs of their free allowances. This is akin to the situation of having no ETS up and running because energy producers do not have an incentive to increase their prices as all their benefits are taxed away and higher prices entail reduced sales and hence lower profits for energy companies: consequently it would reduce the environmental effectiveness of the scheme. A windfall profit tax might also be politically unacceptable, because it would basically force companies to pay taxes equivalent to their windfall profits, while they initially obtained allowances free of charge.

The fourth and final option is to auction the allowances instead of giving them away for free. When the allowances are auctioned by the government, additional out-of-pocket expenses have to be made by power producers. They will then adjust the electricity prices to reflect these

---

[26] Price regulation can give rise to inefficient (over) investment decisions, sometimes described as the 'gold-plated water cooler effect': see Frank (1997) p. 414.

extra costs. Windfall profits will no longer be made. By auctioning the allowances, the financial gain will shift from the energy companies' shareholders to the government. In Europe, electricity consumers considered a carbon mark-up to be more acceptable when electricity producers had to purchase the allowances directly at auction. This is more transparent and easier to understand for consumers. As a consequence, politicians in the EU decided to switch to full auctioning of allowances to electricity companies as of 2013. For some decades, economists have known that free allocation of allowances is important to garner *ex ante* political support for emissions trading from companies.[27] What economists did not foresee is that auctioning allowances could be important in keeping emissions trading politically acceptable *ex post*, namely as a way to overcome consumer protest.

The decision to auction allowances, instead of handing them out for free, does not affect the basic effectiveness or efficiency of the scheme. Effectiveness remains unaltered since the emissions cap does not change by handing out allowances for free or by demanding companies to buy allowances at an auction. Efficiency is guaranteed in the sense that after either free allocation or auctioning of the allowances, companies can then trade their emission rights so that emissions are reduced where this is the most cost-effective to do so. From an efficiency point of view, auctioning does have some additional advantages, which will be examined below.

4. AUCTIONING OF EMISSION RIGHTS

4.1 Advantages of Allowance Auctioning

It is widely recognized in the economic literature that auctioning is an efficient allocation mechanism for the following reasons.[28]

1. Auctions lead to self-revelation of the bidder's private values. An auction mechanism can yield higher revenues than simply quoting a price or repeated negotiations with potential buyers, although bidders are generally reluctant to reveal their preferences because they fear that competitors could take advantage of it.
2. Auctions can be designed in such a way as to ensure 'allocative efficiency': auctions award the allowances to the bidder with the highest valuation for them (assuming that bidders have sufficient

---

[27] Baumol and Oates (1988).

[28] On this issue see also Weishaar (2009) p. 61.

financial resources). However, auctions are capable of demonstrating their superior allocative efficiency properties only if they are tailor-made towards the respective bidders.

3. Prior knowledge of the auction rules provides bidders with a transparent framework of how their bids will be assessed.
4. Compared with the time-consuming negotiations that may occur with an administrative allocation, auctions are relatively fast allocation mechanisms. Nevertheless, the development of an auction mechanism can be a time-consuming process. Also the time and expense required for companies to gather information and to design an optimal bidding strategy can be substantial.
5. What is particularly attractive is that allowance auction revenue can be spent on lowering distortionary taxes, such as labour taxes, for the industry, which economists usually refer to as the 'double dividend hypothesis'.[29] Labour taxes are economically distortionary because their effect on the market price of labour. With such taxes, labour becomes more expensive and thus the demand for labour is less than it would be without the tax. With an emissions trading scheme, using the allowance auction revenue to lower labour taxes is economically desirable as it implies that firms have higher pollution costs and lower labour costs.

4.2 Examples of Allowance Auctions

Notable examples of allowance auctions are those for the US Acid Rain Programme, the RGGI, and the UK, Californian and EU schemes.

The auction revenue for the government can be less than initially foreseen. In California, for instance, demand for allowances at the beginning of 2013 was less than anticipated. As a result, California lowered its forecast for expected allowance auction revenue in 2013 from $1 billion to $200 million. This put the state's plan under pressure to use that revenue for funding a high-speed rail project.[30]

An auction system for greenhouse gas emissions that was employed in the European context prior to the introduction of the EU ETS, successfully circumventing the deficiency of stakeholder support, was the UK Emissions Trading Scheme (UK ETS). It was based on voluntary participation and essentially reversed the auction parties: it was the government that purchased allowance abatement from companies. In the

---

[29] See, e.g., Goulder (1995).

[30] Carbon Market News (2013).

UK scheme auction companies would submit their abatement schedules (that is, how much they were willing to abate for which prices) and the government would buy as much as it could at the auction clearing price. The UK ETS operated in this form from 2002 to 2005.

Within the framework of the EU ETS, allowance auctioning played a minor role between 2005 and 2012 but has become the predominant allocation mechanism since 2013. Auctions are increasingly being phased in, rising from 20 per cent in 2013 to 70 per cent in 2020, while 2027 is the targeted date for full auctioning. With some exceptions for Eastern European Member States, auctioning is the default allocation tool for the power sector. While 88 per cent of the total quantity of allowances to be auctioned will be distributed to Member States, 10 per cent will be allocated for the purpose of 'solidarity' and growth within the EU. The remaining 2 per cent will go to those Member States that, in 2005, were at least 20 per cent below their Kyoto Protocol base-year emissions. Fifty per cent of the auction revenues should (not shall) be used for emission mitigation and adaptation measures. However, sectors and sub-sectors exposed to a significant degree of carbon leakage are eligible to receive up to 100 per cent of their allowances for free. This can be reviewed if an international climate change agreement is ever reached.

4.3 Allowance Auction Design[31]

Auctioning is efficient, but auction design is crucial. An auction format that cannot ensure that the bidder with the highest valuation always receives the item is never allocatively efficient and should therefore not form the basis of an allocation system for emission allowances. Moreover, the essence of auction theory and mechanism design is to enable auctioneers to influence the result of the auction. This is achieved by estimating the 'behavioural parameters' of the likely bidders, such as their level of information, their risk averseness and their willingness to participate in tendering. This assessment is used as a starting point for designing a mechanism that allows for the maximization of revenues.

The legal rules for the auctioning of allowances in an emissions trading scheme need to incorporate various elements, including the bundling of allowances and the timing of auctions.[32] Transaction cost considerations require the government to auction larger lots of allowances, referred to as 'bundles'. This has to be balanced against the desire to allow small and

[31] This section is based partly on Weishaar (2009) Section 3.2.1.

[32] On the design challenges with respect to auctioning greenhouse gas allowances see Weishaar (2008a).

medium-sized enterprises (SMEs) full access to the market. The best access to auctions is granted to SMEs if they can buy low numbers of allowances. This is the case, for example, at the Leipzig Energy Exchange where bidders can purchase single allowances.[33] The transaction costs of purchasing a low volume of allowances at auction will, however, be higher than if they are purchased on the secondary allowance market. In general, it should be observed that the number of bidders that are actually registered with the Leipzig Energy Exchange and the former Blue Next auctions in Paris is limited.[34]

A decision on the timing of auctions has to be made. They could be organized every day, on a monthly or quarterly basis, or clustered in the months prior to the submission of allowances. Timing is relevant, since it may give rise to the phenomenon of a 'declining price anomaly': the empirical phenomenon of a market price which first lies above its true equilibrium and which then falls in subsequent auctions.[35]

The basic choice in auction design is that between:

- sealed-bid uniform-price auctions; and
- auctions that do not provide a uniform price.

These design formats differ in their consequences for efficiency. A sealed-bid uniform-price auction is the chosen format in the Auction Regulation[36] of the EU ETS. Such an auction is 'sealed bid' in the sense that it allows bidders to enter their bids without allowing their rivals the opportunity to respond *ex post* to them. Moreover, such an auction is 'uniform price' in the sense that it requires every bidder to pay the same clearing price. In a sealed-bid uniform-price auction bidders state how many items, in this case emission allowances, they wish to buy at various prices. Bidders then pay for every unit requested at the clearing price.

Uniform-price auctions have three advantages over auctions that do not provide a uniform price.[37] First, every bidder pays the market clearing price, which is equal to the marginal benefits of the emission allowances. Second, less informed bidders are more inclined to participate in this type of auction, which in turn increases liquidity and hence the efficiency of the market. Third, uniform-price auctions that ensure a single market

---

33 See EEX (2011), pp. 6ff.

34 The number of registered bidders in 2012 in Leipzig was 80 and in Paris 115: see below Chapter 6, Section 3.1.

35 Ashenfelter (1989).

36 Commission Regulation (EU) No. 1031/2010 of 12 November 2010.

37 See Milgrom (2004) p. 256.

price are preferred over sequential auctions because the former mitigate the 'price risk' of paying different prices for emission allowances (bidders do not like to pay more than their competitors for identical items).

However, in the presence of market power, auction participants could use 'demand reduction' strategies.[38] In such a strategy a bidder chooses to buy less than it needs in order to lower the clearing price. For the system as a whole, this would be inefficient since the allowances then do not accrue to the bidder with the highest valuation. Large bidders subsequently win too few of the auctioned allowances and small bidders win too many.

In multi-unit auctions dominant players recognize the interdependence of their bidding strategies. A strategy of self-restricting the quantity demanded while bidding a low price can generate a financial benefit for them. For instance, instead of paying a price of $20 for two million units, a successful demand reducer could pay $16 for 1.6 million units. Because demand is reduced, the auction price will be less competitive and result in a lower clearing price. The inefficiency stems from the fact that users which value the item most highly prefer *not* to obtain it at the correct market price.

By contrast, the sealed-bid multi-unit auction format does not suffer from demand reduction. The second-price sealed-bid auction system, referred to as a 'Vickrey auction', is an auction in which bidders submit the prices they want to pay for emission allowances.[39] Bidders win the quantity demanded at the clearing price but pay an amount equal to the highest losing bid for each unit. Given this allocation and payment rule, sincere bidding is a dominant strategy for bidders. The bid submitted determines if the bidder is allowed to buy allowances but does not have an impact on the ultimate price it will have to pay. The price is determined by the highest losing bidder, which is therefore beyond the influence of a successful bidder. Thus, if a bidder wants to obtain the allowances, it must place a bid based upon their true valuation. Consequently, efficiency-distorting demand reduction will not occur.

Sealed-bid uniform-price auctions are likely to be more allocatively efficient than auction schemes that do not have a uniform clearing price, because the costly strategic burden of misjudging the clearing price is not placed upon the bidders. In the presence of market power, demand reduction can be an issue. While this would give rise to a lower auction

---

38 For examples see Weber (1997) and Ausubel and Cramton (2002).

39 Vickrey (1961).

price and hence be detrimental to the auctioneer's proceeds, it would be beneficial to the dominant firm, as well as to small firms. Small firms would benefit from demand reduction since it gives rise to a lower auction price and consequently gives small firms every incentive to participate in the auction. This, in turn, would actually help to mitigate the effectiveness of demand reduction strategies. Auction rules under a sealed-bid uniform-price mechanism are easy for bidders to understand. However, the strategic complexity of placing bids should not be underestimated. Bidders not only have to determine the price they are willing to pay themselves, but also that which their competitors are likely to bid.

In the presence of market power, a Vickrey auction is more efficient since it does not give rise to demand reduction. In the absence of market power, a uniform-price auction is similarly efficient with the additional benefit of simplicity in the auction rules, which may attract more bidders. As a consequence, the sealed-bid uniform-price auction of the EU ETS does not exclude all incentives for market manipulation stemming from demand reduction. If energy companies operate in an oligopolistic market and hold a vast majority of the allowances, such as Germany in the EU, effective monitoring and enforcement mechanisms need to be in place in order to mitigate market distortions.

5. LOBBYING FOR MORE OR CHEAPER EMISSION RIGHTS

Lobbying is likely to occur in every emissions trading scheme in every political system.[40] Every allowance allocation option will be subject to lobbying before it is decided upon. Governments and bureaucrats in the various political cultures are likely to respond differently to lobbying, for instance, depending on the access that entrepreneurs have to officials, the level of democracy, and the level of corruption. Although firms could lobby for a favourable auction design that, for example, excludes reserve prices, in view of the fact that auctioning relies heavily upon competition, in practice lobbying regarding auctions may produce less serious effects. Free allocation of allowances, on the contrary, is much more susceptible to lobbying. Electricity producers and the industrial sectors, for instance, are likely to lobby for a favourable allocation of emission rights.

---

40 See Dijkstra (1999) and Woerdman (2004).

5.1 Lobbying and Inefficiencies

In general, each firm wants to be a seller rather than a buyer of allowances, which means that individual firms (energy companies, for example) have an incentive to lobby for more allowances. This happened in the Californian emissions trading scheme where regulated businesses put pressure on the state to hand out more allowances. The collective result could be over-allocation, as witnessed in the EU ETS in the first trading period 2005–07, with the effect that allowance prices dropped significantly.[41]

Firms that compete in an international product market and use a lot of energy in their production are likely to lobby for cheaper allowances. Because of the emissions trading scheme their industrial production will include the price for greenhouse gases, while they compete with companies from countries where no internalization of environmental costs is required. The former firms could then lobby for some kind of financial compensation from the government and demand that their higher electricity bill is reduced by subsidies.

Lobbying for more or cheaper emission rights has detrimental effects on the efficiency of an emissions trading scheme. In both cases, the social costs of climate damage are not fully internalized. Lobbying for more allowances undermines the scarcity factor and therefore the basis for operating an emissions market. Lobbying for cheaper allowances – for instance, by way of compensation for the increased electricity bill – discourages the use of less and cleaner energy. Finally, where two or more firms lobby, an inherent welfare loss occurs because only one party can 'win' and parties tend to over-invest.[42] On the other hand, there can also be situations where the welfare loss deriving from two 'opposing' lobbyists is less detrimental and closer to the optimum solution than where lobbying occurs 'unopposed'.[43]

5.2 Lobbying in the European Union ETS

The European Union is a clear case in point. 'The EU ETS has developed into a highly complex system, which has certainly been open to corporate influence and partially outright capture'.[44] It has indeed been analysed that industry lobbying behaviour has led to rules and exceptions that

---

41 Jong, Couwenberg and Woerdman (2013).
42 Cullis and Jones (1998) p. 95.
43 Faure (2012) p. 308.
44 Lederer (2012) p. 7.

hinder the efficiency of the trading system.[45] Lobbying has focused not only on more and cheaper allowances, but has also targeted more subtle and sometimes unexpected aspects of the emissions trading scheme, such as the formulation of installation closure rules.

First, industry in the EU lobbied for an inefficiently high number of allowances. Both the electricity producers and the energy-intensive sector lobbied fiercely in the run-up to the establishment of the EU ETS to gain as many allowances as possible. Politicians were concerned that too stringent caps could harm domestic competitiveness. Since the introduction of the EU ETS in 2005 was to have a gradual start to enable parties to gain experience they preferred to be more generous in allocation than to harm industry.

The consequence, however, was over-allocation. On the one hand, lobbying had indirectly contributed to a temporary collapse of the market: the carbon price plummeted to less than €1 in 2007. The number of allowances allocated in the EU as a whole turned out to be 4 per cent more than actual emissions.[46] This could be seen as 'government failure', but on the other hand, nobody could have predicted the exact emissions gap and price drop beforehand. This was made difficult by the inherent price uncertainty in a market for allowances and by incomplete information about the emission levels estimated in advance. Business growth predictions were over-optimistic and turned out to be lower than expected, which meant that actual emissions were also lower than predicted. The European Commission intervened and forced several Member States to reduce their National Allocation Plans (NAPs). As a result, emission levels were set more stringently in the second trading period (2008–12). Because of the economic downturn in 2009, an oversupply situation arose that still persists today. In order to avoid situations of over-allocation in the future, allocation was changed from a 'bottom-up' approach at national level to a 'top-down' approach at EU level (effective since 2013). The European Commission is planning to address the current oversupply situation by 'backloading' (discussed in Chapter 4).

Second, at Member State level industry lobbied for closure rules that would not take allowances away if enterprises ceased to produce or reduced production. Several Member States, however, thought that such allocations should be adjusted during the trading phase and incorporated so-called *ex post* adjustment rules into their NAPs. The European

---

45 Woerdman (2013).

46 Ellerman and Buchner (2006).

Commission, by contrast, thought that such *ex post* adjustments infringed the European rules on allocation of allowances and rejected several NAPs. In an ensuing legal case, the Court upheld the lawfulness of *ex post* adjustments.[47] The practical implications were, however, limited since the NAPs for the second trading period (for Germany, for example) had already been completed.

Interestingly, under the harmonized allocation rules for the third trading period, the Commission does allow *ex post* adjustment in the context of plant closures. Plants that cease to operate lose their allowances in the following year (unless they can start up again within six months); there are also rules which apply to partial cessation of production.[48]

The closure rules make it more attractive to maintain non profit-making production capacity and thus acquire allowances that may consequently be sold. When allowances are highly priced, a firm could hypothetically even invest in production capacity for the sole purpose of acquiring allowances intended for sale.

There are other cases where industry lobbying did not succeed. For many years, metal manufacturers, represented by Eurometaux, and the chemical industry, represented by Cefic, have lobbied for 'performance standard rate (PSR) trading' (or credit trading), where emissions trading is based on an emissions standard per unit of output or energy.[49]

Under the current rules, closing down inefficient plants is basically penalized by a loss of allowances. A more efficient option, therefore, is that, in the event of installation closure or capacity reduction, allowance allocations do not cease but continue to be allocated for the remainder of the multi-year period for which they were allocated. This would create a somewhat stronger incentive to close down old polluting plants.

Third, industry in the EU has lobbied for the inefficient reimbursement of indirect emissions costs. Before explaining the inefficiency of this, it is important first to understand what an efficient situation should look like. As a result of carbon policy, the electricity bill needs to be increased for small and large consumers; as a result of emissions trading this increase will remain limited. In other words, climate policy should, *ceteris paribus*, lead to higher electricity bills since the emission of damaging greenhouse gases should have a price. Energy consumers, including the large ones, are then encouraged to use less and cleaner energy. This is the

47 For a case note on this issue see Weishaar (2008b).
48 See Arts 21 and 22 of Commission Decision 2011/278/EU.
49 Schyns and Loske (2008); see also Van Renssen (2012).

theoretically optimal situation. In practice, however, the energy-intensive industry, such as Eurometaux, the lobby group for metal manufacturers,[50] has lobbied for financial compensation for their increased electricity bill. They fear that, because of international competition, they will incur sales losses if they pass on their climate costs. Industrial production, and thus emissions, could be transferred to countries outside the EU that do not impose emissions caps. Nevertheless, this carbon leakage effect is likely to be limited,[51] for instance, because other cost items weigh more heavily in the relocation decision than the relatively limited CO_2 costs.[52]

In Directive 2009/29/EC (amending Directive 2003/87/EC) compensation for the higher electricity bill is made possible, subject to the condition that a sector stands to lose so many sales that it will only be able to continue production profitably outside Europe. These conditions have recently been developed into new European Commission state aid rules.[53] In the coming years, Member States may reimburse up to 85 per cent initially, and later 75 per cent, of the indirect CO_2 costs to enterprises. These new state aid rules, however, are not efficient. Firms only get into trouble when the allowance price is high – for example, when the price is higher than €20 to €30 per tonne of CO_2. But even at much lower prices of around, say, €5 per tonne of CO_2 it will still be possible for EU Member States to grant a subsidy. EU policy makers thought that, at most, two or three sectors would claim compensation, but in the end no fewer than 15 sectors have done so, including steel, aluminium, copper, zinc, paper and chemicals. The government of the Netherlands, for instance, expects that if the increased electricity bill of these sectors is paid by the state up to the aforementioned maximum, this will cost the country about €100 million a year.[54]

Compensating higher electricity bills for large energy consumers would represent a significant weakening of the EU ETS, since the climate damage is then not sufficiently internalized. Furthermore, unlike countries such as Germany, Finland and Norway, most EU Member States (including the Netherlands) have not set aside money for compensating large energy consumers. If some countries do and others do not compensate, the problem of state aid and distortion in competition arises. Instead of the current regime, it would have been more efficient to allow

[50] Eurometaux (2011).

[51] Mattoo, Subramanian, van der Mensbrugghe and He (2009); Heilmayr and Bradbury (2011).

[52] Sijm et al. (2004).

[53] Communication from the Commission (2012).

[54] Atsma (2012) p. 3.

compensation only when the allowance price is high, leading to a real danger of sales losses and production transfer, and only for those sectors with exceptionally high electricity costs.

Fourth, lobbying is not always predictable and may even have a positive impact on the efficiency of an emissions trading scheme. A clear example of this is the unexpected energy consumer lobby in the EU for the auctioning of allowances. With the political support of Brussels, they contested the so-called 'windfall profits' made largely by the electricity companies on the basis of free allowances. Consumers' electricity bills went up, while the electricity companies had paid nothing for the emission rights. Various politicians and consumer organizations called for change. From an economic point of view, as explained above, these 'windfall profits' are correct: the external costs of emitting greenhouse gases should be internalized and for that reason, *ceteris paribus*, the price of electricity should rise. The allowances might have been acquired for free, but the allowances have opportunity costs if they are used, equal to the carbon market price. 'Windfall profits' are thus not an economic issue, but a political problem. Because free allowances have a market value, shareholders of electricity companies become richer. Consumers found this to be unacceptable.

For the European Commission this consternation created the political space for making a transition to the auctioning of allowances. Economists like auctions because the allowances will accrue to those who value them the highest. However, this transition is contrary to the predictions of economists who considered free allowances to be desirable with a view to acceptance by producers. The regulated industry was confronted with the unforeseen lobbying by consumers who refused to swallow the efficient passing on of the value of allowances acquired for free in the price of electricity.

The case of lobbying in the EU ETS shows at least two things. On the one hand, in the political game of give-and-take the government has, on several occasions, been willing to bow to the wishes of industry, at the expense of the efficiency of emissions trading; this could be referred to as 'government failure'. On the other hand, certainly not all wishes of the industry lobby have been met, such as their desire to completely reshape the EU ETS into a PSR trading scheme. ETS designers must be aware of the lobbying going on both before and during the implementation of an emissions trading scheme. They must also have clear ideas of which design elements are efficient and which would not be efficient in order to steer the political process of designing an emissions trading scheme.

6. CONCLUSION

Before starting emissions trading, an emissions cap or emissions standard must be formulated and adopted. In order to reduce emissions, this cap or standard must be more stringent than current emission levels. That did not happen in the EU ETS, which experienced an ineffective and inefficient situation of over-allocation. Over-allocation of initial allowances is very difficult to remedy and should therefore be avoided by emission trading designers.

The next step is to allocate the emission rights for free or to sell them. In the absence of an emissions cap, free allocation can be based on performance standards. In the presence of a cap-and-trade system, free allocation can be based on grandfathering (allocation based on historical emissions) or on benchmarking (allocation based on clean production standards). Auctioning of emission rights is another option in, among other design variants, sealed-bid uniform-price auctions or in auction formats that do not have a uniform clearing price. Free allocation led to the political problem of 'windfall profits' in the EU ETS: electricity producers and industry passed on the market value of their free allowances to consumers. Although this is economically correct, the consumer lobby contributed to the legal transition to allowance auctioning.

Enterprises are likely to lobby for a favourable allocation of emission rights, but this may have detrimental effects on the efficiency of an emissions trading scheme. In the case of the EU ETS, for instance, lobbying by industry contributed to an inefficient surplus of allowances in the system, to the emergence of inefficient closure rules, and to inadequate compensation of indirect emissions costs (higher electricity bills). Policy makers and ETS designers should not allow themselves to be so easily drawn into regulatory inefficiencies such as those examined above. Instead, they should follow economic theory as closely as possible when designing and implementing emissions trading systems. If it is not possible – as a result of more pressing societal objectives – to follow efficient ETS design solutions, it is important to be aware of the economic costs. Only the careful weighing of the available ETS design and policy options can yield optimal outcomes; finding the relevant information about the costs and benefits of the design choices is essential.

6. Implementation issue 2: The secondary market for emission rights

1. INTRODUCTION

This chapter addresses a crucial issue for the working of any emissions trading system: the market. This area has received less attention in the literature than other aspects of emissions trading design. This neglect is probably attributable to the belief that the emissions trading market will work as does any other market. While not suggesting that the emissions trading market is unique in any way, it still merits attention, since a well-functioning market is the precondition to allow for cost-effective abatement and for realizing the full benefits of an emissions trading system.

Recently, the emissions trading market in Europe has been receiving more attention. After the fraud scandals that occurred in the EU ETS, issues such as market manipulation, market power and market oversight started to attract more interest. All of these issues are considered in the context of the EU ETS in this chapter. First, market manipulation on the spot market is addressed in Section 2; this section emphasizes that market manipulation can also take place in liquid emissions trading markets and should therefore not be underestimated by ETS designers. Section 3 considers market power and oversight issues in the context of EU ETS auctions; this section points out that even the use of auctions does not prevent distortions in competition and manipulation. ETS designers are therefore well advised to carefully consider auction design choices and establish efficient oversight structures. Section 4 reviews documented market irregularities; in particular, it considers criminal activities that can arise in the context of emissions trading systems. Examples examined here include fraud scandals that occurred in the context of the EU ETS, allowance theft and inappropriate behaviour on the part of an EU Member State which, although in compliance with existing laws, involved activities that were capable of undermining the environmental integrity of the EU allowance market. These are critical problems that ETS designers should seek to avoid.

2. SPOT MARKET AND MANIPULATION

This section emphasizes that market manipulation can take place in liquid emissions trading markets. It presents the legal framework applicable to the EU ETS spot market and types of market manipulation that might occur in emissions trading markets.

Unlike the derivatives market for EUAs, which is regulated by EU legislation – namely the Market Abuse Directive (MAD)[1] and the Market in Financial Instruments Directive (MiFID)[2] – the EU ETS spot market was for many years not regulated at the European level. The spot market is decentralized and allows for several trading routes, including various energy exchanges, intermediaries, and via private contracting. It is not transparent and hence difficult to monitor. Supervision of the spot market for EUAs is regulated voluntarily at Member State level, but it does not amount to a coherent or systematic overview of the entire EUA spot market as market supervision exists in only a few Member States. Germany, for example, applies rules on regulated markets to trades in commodities and transactions on exchanges (thus including EUA transactions), while France has devised rules that have made the legal framework for regulatory markets applicable to the spot market.[3]

Pursuant to Article 12(1)(a) of Directive 2003/87/EC (the EU ETS Directive), the Commission, in 2010, examined whether the EU ETS market was sufficiently protected from insider dealing and market manipulation.[4] It concluded that an in-depth study and stakeholder consultation were necessary to examine in greater detail the structure of the carbon market and the level of oversight.[5] Insights from this consultation found their way into the revision and strengthening process of the MAD to enhance the regulation of commodity and commodity derivatives markets to deal with insider dealing and market manipulation. On 20 October 2011, the European Commission published its provisional draft of the amended MAD and the new Market Abuse Regulation

[1] Directive 2003/6/EC of the European Parliament and of the Council of 28 January 2003 on insider dealing and market manipulation (market abuse).

[2] Directive 2004/39/EC of the European Parliament and of the Council of 21 April 2004 on markets in financial instruments amending Council Directives 85/611/EEC and 93/6/EEC and Directive 2000/12/EC of the European Parliament and of the Council and repealing Council Directive 93/22/EEC; Diaz-Rainey, Siems and Ashton (2011) p. 13.

[3] COM (2010) 796 final, p. 10.

[4] See Directive 2003/87/EC, Art. 12(1a), and COM (2010) 796 final.

[5] COM (2010) 796 final, p. 10.

(MAR).[6] The MAR sets out rules and administrative sanctions in relation to insider dealing and market manipulation (market abuse), while the amended MAD now also introduces criminal sanctions. In the third trading period 2013–20, the spot market will fall under EU legislation.

In order to examine if there is a risk of market manipulation in the first place, some descriptive market statistics are useful to consider. The EU ETS has more than 11,000 installations and 6,500 operators, but it cannot be described as a polypoly. The differences in size of these operators are significant. Based on the Community Independent Transaction Log (CITL) data that is available for the years 2005–06, around 10 global ultimate owners (GUOs) accounted for 29.3 per cent of all allowance allocations.[7]

A small number of firms also account for the vast majority of allowance spot transactions. The three largest account holders in our data set – measured in trading volume – account for a very large fraction of all sales and purchases (8.7 per cent for sales and 10.1 per cent for purchases) on the carbon spot market. The following 10 largest companies account for 20 per cent of the sales and 22 per cent of all allowance purchases. The 50 largest companies account for 58.2 per cent of all sales and for 66.5 per cent of all purchases over the considered period of time. As is to be expected, the influence of the global ultimate owners is larger. Here the three largest GUOs – measured in trading volume – account for 11.5 per cent and 13.9 per cent of all sales and purchases. The following eight largest companies account for 20 per cent of the sales and 25 per cent of all allowance purchases. The 50 largest companies account for 67.6 per cent of all sales and for 81.27 per cent of all purchases over the relevant period of time.

From an industrial economics point of view, it is therefore correct to conclude that there is no problem of market power that might give rise to abusive conduct which one might encounter on industrial markets and that would come within the provisions of European competition law. But it bears mentioning that market manipulation does not require dominance in an industrial economic sense in order to occur successfully.

Within the framework of an internal document of the Dutch Ministry of Economic Affairs, Agriculture and Innovation, national EU ETS actors

---

[6] See COM(2011) 656 final, 2011/0298 (COD). At the time of writing, the text is awaiting first reading in the European Parliament. The indicative plenary sitting date for the first reading is 8 October 2013.

[7] Global ultimate owners are the ultimate owners of installations in the EU. Account holders are those companies that hold an account within the EU for trading on the spot market.

were asked about the spot market rules and whether the MiFID rules should be applicable to prevent market manipulation. Interestingly, non-energy companies feared that market manipulation could come from so called non-compliance traders (traders that do not have covered installations). Furthermore, industry organizations lamented the lack of transparency in the market – a point that was also made in a French study on the EU ETS spot market.[8] Traders and members of the energy sector, by contrast, denied the occurrence of market manipulation. This suggests that actors who are more active in allowance trading and may therefore have better knowledge of the market, do not think that market manipulation occurs, while others – being relative outsiders – are in fear of becoming victims of manipulation.

Market manipulation undermines trust in the market and leads to a state in which markets do not respond adequately to changes relating to the manipulation. It is a well-noted phenomenon throughout history. Where assets have been traded, there has been manipulation. The trade of both the United East Indian Company and United West Indian Company, for example, was manipulated by spreading rumours about the cargo being carried on particular ships. It may be evident that, on the basis of such rumours, particular individuals were able to earn a profit at the expense of others. The term 'manipulation' thus has a negative connotation; it describes the abuse of a situation without the counter-party being aware of it.

Several forms of market manipulation can occur in a (spot) emissions trading market; these are presented below in a general way, followed by a description of how they can be used in an EU ETS context.

2.1 Action-based Manipulation

This form of manipulation relates to actions that are undertaken to change the (depreciated) value of a company. In situations where a market actor is forced to take a position and this is known to other market participants, the opportunity exists to take advantage of this. Two scenarios illustrate this point:

1. In a case where a company is in financial distress and a share issue is necessary to improve its financial position, cancelling the issue may not be an option. Market actors can lower the price of the

---

[8] Prada (2010).

stocks shortly before the issue date by creating an artificial oversupply. This negatively affects the issue price of the shares. Market actors can then sign up at a lower issue price. After the shares have been issued and the oversupply is taken away, the stock price should move back to its normal level. This type of price manipulation is risky since the price level does not always restore itself. This could be attributable, for example, to new information indicating that the financial situation of the company is even more serious than expected.

2. In a case where a market actor must sell a substantial block of shares in a company, this can also give rise to oversupply pressure that reduces the market price. Other market participants can then purchase this block of shares at a lower price. After the purchases the stock price should move back to its original level as the short-term oversupply is eased.

These two scenarios show that market participants can employ specific value-relevant transactions for their own advantage, but manipulators themselves can also release information regarding planned sales or purchases of assets. A declaration of intention to effect a (hostile) takeover can have a price-supporting effect. Failure to follow up upon one's declared intentions and the in-time sale of accumulated stocks can yield a profit for the manipulator.

Such action-based strategies can also be used in the context of an emissions trading system. Large traders that enjoy a good understanding of the ETS spot market can take advantage of a high degree of uncertainty on the market and buy allowances. Market actors can try to influence the market in the short term in the following two ways.

First, large companies with the benefit of better information might be able to pre-empt price developments on the spot market by selling allowances shortly before the price falls or buy allowances shortly before the price rises. This type of trading strategy can be described as (market) *timing*.

Second, market actors can influence market expectations by buying large quantities of allowances and thereby bid up market prices. Once prices are high, they can place reverse orders and sell off their EUA holdings. As an alternative, they can sell allowances to lower the prevailing market price and buy them back once the market price has fallen. This trading strategy can be described as taking *opposing positions*.

2.2 Information-based Manipulation

Manipulation based on the release of insider information or the spreading of (untrue) rumours is described as information-based manipulation. Here, prices are manipulated by disseminating misleading information or rumours. The following two examples describe such practices:

1. 'Pump and dump' is a frequently applied form of manipulation based on the spreading of untrue rumours regarding a company. These rumours are used to inflate the market price of a company. The manipulator then sells his stocks when the prices are high. This form of manipulation is, of course, very old, but is particularly relevant in today's internet era when the spreading of (anonymous) information is easier than ever.[9]
2. Manipulation can also take more organized appearances. Aggarwal & Wu[10] describe manipulation regarding the initial public offering of Paravant Computer Systems Inc. In this case the broker participated in keeping supply artificially low in the aftermarket, while at the same time the broker created extra demand. With the strongly increased price in the aftermarket, the market manipulators were able to sell their stocks at a substantial profit.

These information-based attempts to manipulate prices through the spreading of rumours can be very diverse and range from unfavourable profit reports to stating that hazardous substances are contained in a company's products. The distinction between information-based and action-based manipulation is not always clear-cut, since spreading a rumour about a takeover has the same effect on price as an announcement by the party intending to execute such a takeover.

Companies participating in an emissions trading system that enjoy better market insights or insider information can spread rumours that lead other market actors to expect a change in the spot market price or different market demand or supply conditions. Besides information-based

---

9 See, for example, the case before the Amsterdam Court (Rb Amsterdam, 3rd July 2003, JOR 2003/204, nt. Italianer). In this case an internet discussion forum was used to suggest that a significant package of shares of Cardio Control were sold and that this was the start of a big price rise. Since it could not be established that the defendant actually benefited from the manipulation, he was convicted only for the attempted manipulation. Other examples are described by Leinweber and Madhaven (2001).

10 Aggarwal and Wu (2006) p. 1934.

abuses, companies can use signalling strategies to manipulate the market. As described above, signalling strategies frequently occur in an opaque form that does not disclose the identity of the signalling party or the true source of information.

2.3 Trade-based Manipulation

Trade-based manipulation is a form of manipulation involving the buying or selling of assets without a specific transaction which can be identified to impact on the market price of the company, but which still gives rise to misleading information. This is a residual category and includes actions to manipulate prices by means of double transactions. Traders may find it lucrative to manipulate the market by first placing an order and shortly afterwards placing an opposing order to influence the price level as well as to influence (volume) statistics. Market actors are thereby able to influence information that other market participants use for their decision-making. In addition, short sellers can be cornered (market squeeze) by taking substantial positions. As a result of the resulting relative scarcity, short sellers may be forced to close their positions against inflated prices.[11]

Companies operating on the EU ETS spot market can try to use trade-based manipulation strategies to squeeze the market by taking a particular position on the derivatives market, requiring sellers to deliver EUAs in the future, while bidding up the EUA price on the spot market before the delivery date. In this way manipulators are able to earn a profit by selling off their EUAs holdings at a higher price.

2.4 Concluding Remarks

Even though there is no hard evidence that market manipulation is taking place within the framework of the EU ETS secondary market, it still remains a theoretical possibility. Given the very low price of EU allowances at present, the possibility that the European carbon market could become a target for manipulative activities appears to be remote. Other emissions trading systems that are succeeding in generating a

[11] A well-known example of market cornering concerns the practice of the Salomon brothers in the 1990 and 1991 sales of Treasury Notes. By purchasing a significant part of the notes auctioned by the Treasury, short sellers were unable to close their books and were forced to trade at very unfavourable conditions with Salomon traders. See Jordan and Jordan (1996).

strong positive market price, however, should pay due attention to manipulation in their emissions trading design.

3. AUCTIONING AND MANIPULATION

This section examines how situations of market manipulation can arise in the auction context of an emissions trading system. Aspects of efficient oversight are found to be vitally important in enabling the competent authorities to protect the emissions trading market.[12]

As in the previous section, the EU ETS spot market initially was not subject to market oversight; market manipulation, therefore, could not be detected by public enforcement agencies. Concern over the possibility of market manipulation has led to legal reforms and, in the third trading phase, the EU ETS spot market is governed by EU legislation.

From both an economic and a policy perspective, auctions are attractive allocation mechanisms. They lead to the self-revelation of the bidder's valuation of emission allowances and they can be designed in such a way as to ensure allocative efficiency. Furthermore, auctions legitimize transfers – of sales in the absence of market prices – which would otherwise be suspect. Because time-consuming negotiation is not required, auctions are fast allocation mechanisms. They are attractive from a policy perspective since they allow the raising of funds that can be used to reduce distortionary, and thus welfare-reducing, taxes (double dividend hypothesis), and avoid politically contentious windfall profits. Probably because auctions constitute the epitome of a market-based tool, there has been little attention in the literature on the market forces that could undermine emissions trading auctions.

The European Auction Regulation (Commission Regulation 1031/2010/EU) addresses a number of legal and economic issues such as market access, size of lots, withdrawal and timing of auctions. The legislator has also included provisions that address market abuse.

Because different terms have different meanings in different social science disciplines, it is useful to describe market abuse in more detail. Within the framework of Directive 2003/6/EC on insider dealing and market manipulation (market abuse), abuse is used to describe both insider dealing and market manipulation. It is also with this meaning that it is used in the Auction Regulation. Note that in the context of industrial economics (the area of economics dealing with cartels and abuse of

[12] A version of this section is published in Weishaar and Woerdman (2012).

dominance) the term 'abuse' requires the firm to have 'market power' (dominance) in the market. In the context of auction design, abuse seems to play a lesser role. Nonetheless, for in-auction strategies to be able to manipulate prices via demand reduction and signalling, the size of a bidder appears to be a factor that contributes to the success of the strategy. When we use the term 'market power' in this section we do not refer to the possibility of raising prices on the industrial markets, but to being a large buyer in the allowance auction.

Two types of market manipulation can arise in relation to allowance auctions. These are demand reduction and signalling. *Demand reduction* describes the situation where large firms understate their demand for allowances so as to be able to buy allowances at lower prices. *Signalling* describes practices that can be used by firms to negotiate the price in the context of an allowance auction. Such strategies can arise not only during the auction, but also before the auction takes place. In line with the Auction Regulation, we use the term 'market manipulation' to indicate that we are not considering insider trading but only in-auction demand reduction and signalling, insider trading being beyond the scope of an auction mechanism design.

This section will examine whether the legislator has duly addressed the particular forms of market manipulation that can arise in the auction context. In doing so, we first use general auction-theoretic insights to evaluate the legislature's specific design choices for the auction mechanism in light of the regulatory objective to prevent market abuse (Section 3.1). Because the chosen auction design does not prevent market manipulation, we then examine the legal framework to determine whether the provisions on enforcement and monitoring are able to address market abuse efficiently (Section 3.2).

3.1 Market Manipulation and Auction Design

This section examines whether the design of the auction mechanism, as described in the EU Auction Regulation, is effective in preventing market manipulation. The design choice is contrasted with an alternative auction design choice.[13]

Allowance auctions are subject to the Auction Regulation. In the Regulation the legislator has opted for a sealed-bid uniform-price auction, which prevents bidders from responding *ex post* to their rivals. The

---

[13] For a more detailed treatment of auction theory in the context of the EU ETS see Weishaar (2009).

chosen auction method means that every bidder pays the same clearing price. From an allocative efficiency point of view, auctions that ensure a single market price are preferred over sequential auctions, because they mitigate the price risk of paying too much for the same item.[14] In general, an auction format that cannot ensure that the bidder with the highest valuation always receives the item is never allocatively efficient, and hence cannot be a viable first choice for an allocation system for CO_2 emission allowances.[15]

The auction mechanism selected by the European legislator effectively addresses signalling that can occur during an auction. A sealed-bid auction does not enable bidders to 'negotiate' bidding solutions with each other during the actual auction process. With respect to this type of signalling the legislator has made a good design choice. Signalling that occurs on the EU ETS market before the actual start of the auction, of course, falls beyond the scope of auction mechanism design and is addressed below.

In the presence of market power, however, the auction format selected by the European legislator is susceptible to demand reduction. With this strategy, a bidder chooses to buy less than he needs in order to have an effect on the clearing price. If successful, this leads to inefficiency[16] since goods do not then accrue to the bidder with the highest valuation for them.[17] Large bidders subsequently win too few allowances and small bidders win too many. Economists refer to this as allocative inefficiency. The inefficiency stems from the fact that bidders who value the item most highly in fact prefer *not* to obtain it at the correct market price.

Another auction mechanism, the Vickrey auction, would prevent demand reduction strategies. With the Vickrey auction, bidders submit prices for emission allowances; they win the quantity demanded at the clearing price, but only pay an amount equal to the highest losing bid. Given this allocation and payment rule, sincere bidding is a dominant strategy for bidders. The bid submitted determines whether the bidder is allowed to buy the goods, but does not have an impact on the ultimate price he has to pay. The price is determined by the highest losing bidder

---

14 Milgrom (2004) p. 256.

15 In that case, allocative efficiency would then have to be established on an efficiently operating secondary market.

16 Inefficiency is created by differential bid shading, i.e. when bidders with identical marginal values reduce their bids by different amounts so that awarding the bidder who values the item most is impossible: see Ausubel and Cramton (2002) p. 4.

17 For examples, see Weber (1997), as well as Ausubel and Cramton (2002).

and is therefore beyond the influence of a successful bidder. Thus, if a bidder wishes to obtain allowances, he must place a bid based upon his true valuation. Consequently, the demand reduction that undermines allocative efficiency will not occur. Since the Vickrey auction is also a sealed-bid auction mechanism, it does not allow for any exchange of information at auction and therefore prevents in-auction signalling. In that respect it performs as well as the European auction mechanism.

It can therefore be concluded that, with regard to signalling, both auction types are equally effective. In the presence of market power, however, the Vickrey auction is more efficient as it avoids demand reduction. In the absence of market power a uniform-price auction is equally efficient, but has the additional appeal of simplicity, which may attract small bidders. Which auction format is more desirable, therefore, will depend on the question of whether there is market power in the auction market. At least in the second trading period, in which 10 per cent of the allowances in the EU ETS were auctioned, only a few enterprises took part in the auctions. The number of participants registered for auctioning on the two largest spot exchange platforms in 2012 was 80 at the Leipzig EEX[18] and 115 at Blue Next[19] in Paris. Thus, participation in the auction market appears to be limited to a small number of bidders. This suggests that most of the 5,000 operators of the EU ETS are relying more on the secondary market and on the derivatives market than on the auction market, even though a simple auction mechanism is used. The low number of bidders at auction suggests – all else being equal – a lower level of competitive pressure. Because energy companies hold the vast majority of allowances in several Member States and operate in an oligopolistic market, effective monitoring and enforcement mechanisms need to be in place in order to prevent market distortions.

3.2 Auction Regulation and Market Manipulation

This sub-section examines whether the provisions of the Auction Regulation on enforcement and monitoring are sufficiently well designed to address market manipulation efficiently. Enforcement issues related to the following three issues are analysed below: (i) concurring competences of different enforcement entities; (ii) potential inefficiencies in enforcement

---

18 See http://www.eex.com/en/document/87099/Emissionen_englisch.pdf.

19 See http://www.bluenext.eu.

schedules; and (iii) the delicate balance between public and private enforcement.

Inefficiencies from concurring competences

When several enforcement bodies are charged with closely related tasks, inefficiencies can arise from a number of causes: for example, duplication of work; institutions refraining from taking action because they expect another authority to act; turf wars between regulatory agencies where two or more agencies seek to dominate one regulatory field at the same time; incomplete oversight and incomplete flows of information. The Auction Regulation assigns tasks to several entities which contribute in different ways to containing market manipulation. These bodies are the three national auction platforms to be designated by the United Kingdom, Poland and Germany; the auction platform that will be selected in an EU-wide public procurement procedure; an Auction Monitor; several national bodies supervising the financial sectors;[20] and the European Commission.

Even though the large number of actors could give rise to concurring competences, the European legislator has tried to mitigate some of the associated inefficiencies by its choice of legislative instruments. While the Auction Regulation takes the form of a regulation, and thus has general application and is binding in its entirety,[21] a number of the relevant enforcement provisions refer to directives that grant Member States greater discretion in implementation.[22] The seemingly concurrent competences to address market abuse by bidders in violating Directive 2004/39/EC[23] on financial instruments on the one side, and the tasks of the competent authority to address market abuse under Directive 2003/6/EC[24] on insider dealing and market manipulation on the other, could be avoided if Member States were to clearly delineate the competences of the competent national authorities. Member States could also determine

---

[20] These are the competent national authorities charged with the tasks under Directive 2004/39/EC on financial instruments, Directive 2003/6/EC on insider dealing and market manipulation (market abuse) and Directive 2005/60/EC on the prevention of the use of the financial system for the purpose of money laundering and terrorist financing. The work of authorities designated under the latter Directive are not considered here.

[21] Treaty on the Functioning of the European Union, Art. 288.

[22] In particular, references are made to Directive 2004/39/EC on financial instruments and Directive 2003/6/EC on insider dealing and market manipulation (market abuse).

[23] See Directive 2004/39/EC, Art. 50.

[24] See Directive 2003/6/EC, Art. 12.

that one single competent national authority is charged with implementing the relevant provisions of both Directives. In this way potential inefficiencies stemming from both authorities taking action, and no one authority feeling responsible to take immediate action, could be avoided.

One should also examine whether the concurrent competences give rise to an enforcement structure in which one organization has complete oversight over the market. In this context the flow of information between authorities is crucial. In order to determine if any entity has the ability to ensure detection of market manipulation, it is not only information gaps between the various auction platforms and competent national authorities that need to be overcome; the enforcement authority must also have access to information relating to the auction market and the secondary market for emission allowances, as well as information concerning the covered industrial sectors themselves.

The competent national authorities obtain their information from the various auction platforms,[25] from other national authorities,[26] and from bidders.[27] The legislator has thus provided ample obligations mandating the exchange of information in order to facilitate cooperation between any of the entities involved. For the purpose of drafting its reports,[28] the Auction Monitor is informed by the auction platforms of any suspected market manipulation and measures they have undertaken.[29] The Auction Monitor receives the information on suspected market manipulation directly from all four auction platforms: entities that deal closely with the conduct of the bidders themselves. As a consequence, the Auction Monitor may be expected to have a good overview of the auction market.

Yet it may be questioned whether the Auction Monitor is an entity that is well designed to effectively detect market manipulation. The Auction Monitor receives information on the auctions and, upon request, also from competent national authorities,[30] but it may not have knowledge of secondary market transactions and on the specific demand patterns of

---

[25] See Auction Regulation, Art. 56.

[26] Confidentiality of information does not constitute a bar for the transmission of information: see Auction Regulation, Art. 62(6).

[27] Ibid., Art. 42(5).

[28] The Auction Monitor has the task of drafting a report on each auction, a consolidated annual report and specialized reports at the request of the Commission and Member States and on its own initiative in the case of a breach of the Auction Regulation (see Auction Regulation, Art. 25).

[29] Auction Regulation, Art. 56(2).

[30] Ibid., Art. 53.

companies. This information is essential for detecting if a company employs strategies of demand reduction.

The Auction Monitor may also be expected not to be able to closely monitor the announcements of companies and their trade associations regarding their particular business outlook in the various Member States. This implies that the Auction Monitor does not have the necessary information on bidders' demand, nor on business prospects and announcements, which is crucial to determine whether bidders buy fewer allowances than they need or whether they are engaging in misleading signalling tactics.

It should also be observed that the secondary market for emission allowances is decentralized. Various trading possibilities are available: exchanges can occur via energy exchanges, through intermediaries and private contracting, and encompass both allowances and derivatives. At present, only derivatives fall within the ambit of the MiFID;[31] this makes attempts to effectively monitor firms' behaviour challenging, although under the most recent Commission proposal for the revision of the MiFID emission allowances will also be covered.[32] It should therefore be questioned whether the Auction Monitor, as the entity charged with monitoring market manipulation in the EU ETS auctions, should not be linked more closely with the competent national authorities, especially with those designated under Directive 2003/6/EC, as well as with the national competition authorities. It may be asked how large-scale market abuses are detected in ways other than direct red-flagging by one of the auction platforms in cases where, for instance, the auction price was found to be unduly low.[33]

Inefficiencies in enforcement schedules

Under the Auction Regulation, the most extensive powers of enforcement against market manipulation are vested with the competent national authorities. Auction platforms are, however, permitted to undertake

---

[31] Diaz-Rainey, Siems and Ashton (2011) p. 13.

[32] See COM(2011) 656 final, 2011/0298 (COD), Annex I, section C, point 4. The proposal is presently with the European Parliament for the first reading. The indicative date for the first reading has been set for 8 October 2013.

[33] See Auction Regulation, Arts 7(6) and 7(7). Auction platforms are also mandated to adopt structural provisions aimed at preventing and detecting market manipulations (Art. 42(2)), although here, too, the information processed can be expected to draw only from the actual auction platform and not include data on the actual demands of undertakings.

independent action without having to consult national authorities.[34] Competent national authorities under Directive 2003/6/EC and Directive 2004/39/EC have similar enforcement powers. Pursuant to the Directives these authorities are to be vested with at least the following rights: (a) to require the cessation of any practice contravening the Directive; (b) to order the suspension of trading in the instrument; (c) to order the freezing and/or sequestration of assets; and (d) the temporary prohibition of professional activity.[35] Furthermore, criminal penalties and administrative law sanctions could be used under national law to compel compliance. Of course, the competent authorities are also endowed with access to information and investigative powers.

Yet not only the competent national authorities have the authority to take action to address market manipulation, but the auction platforms themselves also have the right to introduce measures. These instruments include the setting of a maximum bid size and the taking of any other remedial measure necessary to effectively mitigate an actual or potential discernible risk of market manipulation, after consulting the Commission and obtaining its opinion.[36] Member States and the Auction Monitor may be consulted by the Commission. Action taken by the auction platforms is subject to the extra-judicial complaint mechanism as well as to the right of appeal.[37]

It may thus be wondered how an efficient enforcement schedule that ensures an optimal degree of deterrence could be developed between national legal systems and the four auction platforms if the latter are able to take measures independently of the influence of the competent national authorities. The coherent and effective enforcement of rules to prevent abuse in an EU ETS auction context may not be achievable. On a positive note, however, it is encouraging to observe that the measures taken by the auction platforms must be subject to a dispute settlement system which also offers access to judicial review. This may help to mitigate error costs and lead to an *ex post* correction of excessive measures.

Inefficiencies in the mix of public and private enforcement

The enforcement of market manipulation under the Auction Regulation takes the exclusive form of public enforcement. The role of private parties is restricted to the obligation to provide the public enforcement

---

34 Ibid., Art. 57.

35 Directive 2003/6/EC, Art. 12; Directive 2004/39/EC, Art. 50.

36 Auction Regulation, Art. 57.

37 Ibid., Art. 64.

authorities with information on perceived market manipulation.[38] Generally this would be evaluated negatively since important insights and enforcement incentives are not used to deter market manipulation. However, in the context of the Auction Regulation, a more nuanced conclusion must be drawn if one analyses not only in-auction demand reductions but also signalling taking place on the secondary market.

Let us first consider in-auction demand reductions. Unlike the situation of an industrial market setting where firms subject to manipulation are likely to suffer loss and therefore have an incentive to take legal action, this does not seem to be the same for EU ETS auctions. Undertakings participating in an allowance auction that is subject to demand reduction, on the contrary, are likely to benefit themselves. As in the case of the undertaking engaging in demand reduction, competing bidders will also pay a lower auction price. In such a context they do not have a direct incentive to provide the auction platform with information about the market manipulation. The legal obligation to notify the competent national authority about a suspected manipulation must therefore be subject not only to a penalty but also to monitoring, since otherwise it cannot be effective. In any event the enforcement of this obligation may be very difficult in practice.

While the above appears to be true with regard to demand reduction, the same may not be the case with market manipulation that takes the form of signalling on the secondary market and does thus not occur within the auction setting itself. An undertaking planning to sell emission allowances on the secondary market may seek to induce market participants to believe that the price of the allowances will increase. Changes in expectations will not only have repercussions on the secondary market but also on the auctioning market. On the secondary market as well as on the auction market, participants would pay too much for their emission allowances. In this context, private undertakings would incur loss from market manipulation and a role for private actors in the enforcement process would be helpful. In addition to an obligation to notify the competent national authorities about suspected market manipulation, a positive incentive structure of allowing undertakings to claim damages for loss incurred may be efficient, particularly in situations where public authorities are unable or unwilling to initiate investigations. The possibility of recovering damages does not appear to be incorporated in the Auction Regulation, but could stem from private enforcement claims under provisions of European competition law or national law.

---

[38] Ibid., Art. 42(5).

While private enforcement stemming from competing allowance bidders may not be successful without elaborate incentive schemes in the context of demand reduction, private enforcement may be an important enforcement ally in the context of signalling that occurs on the secondary market. Incentive structures encouraging private parties to take an active role in detecting and prosecuting market manipulation should therefore be ensured.

3.3 Concluding Remarks

This section has shown that if the chosen auction design does not prevent all forms of market manipulation, there is an important role for the legal framework to ensure efficient monitoring and enforcement to prevent such manipulation. ETS designers should therefore pay due attention to market oversight structures. With regard to the European legislation several inefficiencies can be identified. These relate to concurring competences of enforcement agents, inconsistencies in enforcement schedules, as well as imbalances in the mix of public and private enforcement. Some of the inefficiencies may be avoided at Member State level where the laws are implemented. It would appear useful for designers of emissions trading systems to take into account the challenges discovered in the European context and try to avoid them.

4. DOCUMENTED MARKET IRREGULARITIES

The EU emissions trading market is a huge market. Of course, its market size varies together with the value of its emission allowances. Also, as the fraud cases have shown, even at low market prices for emission allowances it is clear that the value of the EU ETS is attractive enough to attract the attention of criminal minds. Once the criminal business models have been developed, they could find new applications in other emissions trading schemes. ETS designers should therefore pay close attention to the EU experience. The EU ETS has suffered from several kinds of fraudulent behaviour, criminal and non-criminal. In this section three types of market irregularity are reviewed. These include value added tax (VAT) fraud, emission allowance theft and undue Member State behaviour. Each is treated in turn.

4.1 VAT Fraud

Fraud regarding VAT typically concerns high-value low-volume goods. These are goods that are easily transported and promise a decent profit. Emission allowances can be transferred within the EU within a few minutes and, given the liquidity of the market, can be easily bought and sold. The VAT fraud in European emission allowances occurred between the summer of 2008 and 2009.

To instigate a VAT fraud, fraudsters take advantage of the fact that VAT has to be paid when purchasing goods within a country, but is not payable if those goods are exported. Participants in the EU ETS fraud took advantage of this by purchasing emission allowances in, for example, the Netherlands, without paying VAT, and selling them on to a purchaser in France without surrendering the VAT that was due on that purchase to the French tax authorities. This type of activity is commonly referred to as 'missing trader intra-community' (MTIC). In order to earn additional funds, fraudsters could purchase allowances in France and export those again to the Netherlands. If the fraudster paid VAT (or claimed to have done so) for the emission allowances in France, he can ask for a VAT refund from the French Finance Ministry when exporting the emission allowances. This carousel can start again in other countries.

Besides of the obvious damage to a country's finances, such behaviour is believed to impact on trading volumes and thereby distort prices. Nield and Pereira[39] report that trading volumes at the largest European spot exchange for EUAs – Bluenext, located in Paris – were inflated by over 85 per cent during the high time of the EU emissions allowances fraud.

As a consequence of the VAT fraud, French authorities had to discontinue trading and excluded emission allowances from VAT. An alternative way to prevent VAT fraud was chosen by the Netherlands. The Dutch authorities applied a reverse-charge system in which it was not the seller but the purchaser of allowances who had to pay the VAT to the public authorities. The revised VAT Directive[40] granted Member States the power to apply this option in order to prevent VAT fraud on the EUA market.

---

[39] Nield and Pereira (2011).

[40] Council Directive 2010/23/EU of 16 March 2010; see also Council Directive 2013/43/EU of 22 July 2013.

4.2 Theft

Another problem for the EU ETS emission allowance market occurred in the form of several instances of EU allowance thefts. Starting in January 2010, fraudsters posed as registry administrators to induce unsuspecting victims to register their account details. The details were then used to steal allowances from the victims' accounts. These so-called 'phishing' attacks occurred on several occasions – attacks were reported in respect of Germany, Romania, Italy, Austria, the Czech Republic and Greece – and ultimately led to the interruption of the entire trading on the EU ETS market on 19 January 2011. National registries were allowed to commence trade again if they could prove to the European Commission that they had at least minimum security standards in place to prevent further phishing attacks.

Even though the amount of stolen emission allowances was very low (around 0.003 per cent of the EUA allowances outstanding) and many allowances could be returned to their rightful owners, the theft created considerable problems. Despite the clear identification of each and every EUA, not all allowances could be recovered.

Consequently, purchasers of allowances were unable to determine if the allowances they had bought, or were holding, were stolen or genuine. Given the different property rights regimes in the Member States, there was uncertainty over whether purchasers were indeed holding title to the allowances they had purchased. The market soon reacted to provide detection tools and trading conditions under which companies were offered the more secure purchases they sought. Despite these developments, transaction levels on the spot market fell significantly, remained low for a considerable period and were then translated to the derivatives market, which priced potential legal risks into the market price of futures.[41]

4.3 Nation State Creativity

A further shock that undermined market certainty and consumer confidence in the EU ETS market occurred in March 2010, when the Hungarian government converted 1.7 million certified emission reductions (CERs), which had already been surrendered by national installations, into assigned amount units (AAU). These AAUs were then legitimately sold by the Hungarian government. The risks for EU ETS

---

[41] See Nield and Pereira (2011).

compliance buyers and traders arose from those AAUs that were sold back onto the EU market. Those allowances that had already been surrendered for compliance purposes could, of course, not be used again for compliance within the EU ETS framework. So purchasers of those AAUs were afraid to buy units that they would not have been able to use for their compliance purposes. The actions of the Hungarian government complied with European law, but they still damaged the reputation of the EU ETS and gave rise to double counting. An amendment of the European rules has effectively closed this legal gap. Emissions trading designers should therefore pay close attention to the legal details when they establish ties with other jurisdictions.

4.4 Concluding Remarks

The frauds and thefts of emission allowances have not impaired the integrity of the EU ETS, although they have caused economic loss and discredited the EU ETS in the eyes of the general public, policy makers, and possibly negotiators at international climate change forums.

Completely preventing fraud and theft of allowances may prove to be impossible. The centralized Commission Transaction Log that is used in the third trading phase of the EU ETS may be better protected than national accounts, but it may also constitute a more lucrative target and hence lead to more cyber attacks.

If the ETS to be designed includes several jurisdictions or will be linked with other trading schemes, VAT tax refunds could also arise outside the EU context. ETS designers should therefore consider how to prevent such frauds. One way to achieve this is to exempt emission allowances from VAT or to have the purchaser of allowances pay the tax.

In order to avoid such illicit behaviour, ETS designers should set stringent security measures to discourage fraudsters from opening trading accounts. They could also consider the introduction of a mandatory time gap between sales of allowances. This would help to reduce the speed with which stolen allowances can be dispersed over the trading system. Moreover, they should put centralized response protocols in place to prevent uncoordinated actions being taken by the various jurisdictions involved in the emissions trading scheme. Furthermore, ETS designers should clarify the legal nature of the property rights in emission allowances so that the legal treatment of stolen allowances is uniform and predictable. The harmonization of criminal laws and penalties for the theft of allowances may facilitate deterrence of cyber attacks between different jurisdictions. As can be seen from the above examples, market participants are in need of protection. It is the task of ETS designers to provide that protection.

5. CONCLUSION

This section has examined an easily underrated issue in ETS design: the market. In the context of emissions trading the market is usually evaluated very positively as an instrument that generates efficient outcomes in the presence of uncertainty and information asymmetry between the regulator and covered entities. It is perhaps for this reason that the realization has only recently dawned on policy makers and academics alike that ETS markets are not so very different from other markets. They need tighter rules to operate effectively.

This chapter focused on the EU spot market and pointed out that even markets as large as the EU ETS could be subject to manipulation. There is at present no hard evidence for this, and the current low allowance prices do not make it a very attractive market to manipulate, but it remains a possibility. When formulating market oversight structures, ETS designers should take such problems into account. The EU took a very long time to propose laws to address market manipulation. ETS designers should learn from this experience and establish a sound system from the beginning.

Section 3 of this chapter examined the EU ETS auctions for their susceptibility to manipulation and abuse. It was found that the chosen auction design is simple and, in principle, should be able to attract a large number of bidders. In practice, this is not the case. Only a very few operators are actually taking part in allowance auctions. Because of this, it is possible that only a few bidders are present at auctions, and large bidders may be able to take advantage of the situation and 'demand reduce' in order to pay lower prices. This possibility is not addressed by the chosen mechanism design, which should then translate into the need to have effective oversight structures in place to avoid such abuses. In the case of the EU ETS auction system, we found that there remain a number of deficiencies relating to concurring competences, enforcement schedules and the mix of public and private enforcement. ETS designers are therefore encouraged to devote more attention to the technical area of effective and efficient market oversight.

As a final element, this section examined the recent fraud cases that undermined the credibility of the allowance market. Here ETS designers will need to take adequate precautions against theft and fraud. They will also need to establish clear rules on how their fellow ETS partners deal with the surrendered allowances they receive. Any legal lacuna in this area will undermine the environmental effectiveness of the ETS being designed.

In light of the above, we conclude that the EU ETS offers system designers a productive field of study. Problems relating to the secondary market should not be disregarded or taken lightly. It is, after all, the task of emissions trading designers to shape the legal framework necessary to allow the emissions trading market to prosper and to ensure that it can reduce greenhouse gas emissions in a cost-efficient way.

7. Implementation issue 3: Operational aspects of emission rights

1. INTRODUCTION

Emissions trading systems consist of rules relating to initial allocation, trading and surrender. But this is only part of the story. There are other areas that ETS designers need to consider since an emissions trading system needs to be based on a wider framework of administrative rules and systems that allow it to function in the first place. This chapter addresses issues arising in the context of those systems that enable an ETS to operate. Elements considered here relate to monitoring, reporting and verification (Section 2), state-owned enterprises (Section 3) and transaction logs (Section 4). Each is treated in turn.

2. MONITORING, REPORTING AND VERIFICATION

Monitoring, reporting and verification (MRV) are at the heart of the operation of any emissions trading system and a precondition for its functioning. Emissions trading systems are based on the surrendering of emission allowances in accordance with the particular obligations of individual emitters. In order to determine how many allowances must be surrendered, information on actual emissions is needed. This information is provided by the MRV process. Non-compliance with the surrender obligations generally results in sanctions that are capable of incentivizing future compliance.

It goes without saying that, in order to allow for an orderly functioning of the emissions trading market, the market actors must have trust and confidence in its operational processes. MRV rules therefore respect important legal principles such as completeness, consistency, transparency, trust and truthfulness.[1] Covered entities will generally be

[1] See Jakob-Gallman (2011) p. 81.

obliged to ensure that the determination of emissions is neither systematically nor knowingly inaccurate. Due diligence will be required in order to ensure that the calculation and measurement of emissions exhibit the highest achievable accuracy.[2] Covered entities will also be obliged to facilitate the reasonable assurance of the integrity of emissions data to be reported and be obliged to support the determination of emissions by using appropriate monitoring methodologies.

An integral part of MRV is the actual monitoring. MRV rules will therefore prescribe permitted monitoring methodologies and specific sectoral rules. The EU ETS is a clear example of this. Here, there are two types of methodology for stationary installations. One is calculation-based; the other one is measurement-based. The calculation-based methodology determines emissions from source streams – that is, from specific fuel types, raw materials or products containing carbon. Combustion emissions are calculated by multiplying the activity data related to the amount of fuel combusted.[3] The measurement-based methodology determines emissions by means of a continuous measurement system.[4]

Operators must consider all relevant aspects, including the location of the measurement equipment, its calibration and measurement, quality assurance and quality control.[5] Measurement-based emissions must be corroborated by a calculation-based methodology.[6] In addition, there are rules that require the measurements to be carried out according to particular specifications and by accredited laboratories. Of course, there are specific rules governing data aggregation and how to deal with situations of missing data.

For the aviation sector in the EU ETS there are specific monitoring methodologies. Each aircraft operator determines its annual CO_2 emissions by multiplying the annual consumption of fuel by the relevant emissions factor. In this way, the fuel consumption for each flight and other consumed fuel, including fuel consumed by auxiliary power units, is measured.[7]

It is crucial that each emitter is obliged to monitor emissions based on a monitoring plan that takes into account the nature and functioning of

---

[2] See Art. 7 of Commission Regulation (EU) No. 601/2012 of 21 June 2012 on the monitoring and reporting of greenhouse gas emissions pursuant to Directive 2003/87/EC of the European Parliament and of the Council.

[3] Ibid., Section 2.

[4] Ibid., Section 3.

[5] Ibid., Section 3, Art. 42.

[6] Ibid., Art. 46.

[7] Ibid., Chapter IV, Art. 52.

the emitting installation or means of transportation, and must be approved by the competent authority.[8] The monitoring plan consists of detailed, complete and transparent documentation of the applied methodologies and is supplemented by a risk assessment.[9]

There are also data management and control requirements that must be established. Emitters are obliged to establish, document, implement and maintain written procedures for data flow activities for the monitoring and reporting of greenhouse gas emissions. They must ensure that the annual emissions report resulting from data flow activities does not contain misstatements and conforms with the monitoring plan, the written procedures and Commission Regulation 601/2012.[10] Similarly, emitters are responsible for quality assurance and are obliged to maintain a quality control system.[11]

Emissions reporting must be verified. Verification constitutes an effective and reliable tool to assure the quality of the control procedures, and should serve to improve both the monitoring and reporting of emissions.[12] Verifiers need to be accredited in order to ensure that the verification of reports is conducted by entities with command of the necessary technical competence to perform the tasks in an independent and impartial manner in full conformity with the regulation.

One major problem of MRV is the accuracy of the data. In particular, if several actors or even several jurisdictions are involved in the process, differences in MRV practices can lead to more or less stringent obligations for emitting entities. This, in turn, can lead to distortions in competition.[13] Given that MRV is a very technical and complex area, the devil will frequently be found in the detail. Equal entities must be treated and evaluated equally. If this is not safeguarded, emissions trading may not be able to deliver cost-effective abatement. The reason for this is that the relative position of operators (whether they are long or short) is distorted. Trading will help to cover the position. But if the position is incorrectly established, it will lead to erroneous decisions concerning abatement. The consequence is societal loss.

In comparison with command and control instruments, emissions trading empowers covered entities. It is not surprising, therefore, that the MRV process also affords a pivotal role for the covered entities. This

---

[8] Ibid., Art. 11.
[9] Ibid., Art. 12 and Annex I.
[10] Ibid., Art. 57.
[11] Ibid., Arts 58 and 59.
[12] Ibid., Art. 6.
[13] Zwingmann (2007) p. 140, but also Peeters (2006) p. 183.

strong focus on self-monitoring and self-reporting, however, requires a well-developed system of inspections and enforcement. It necessitates a system that may, perhaps, be even more stringent than is the case under comparable command and control instruments.[14] Yet it should also be pointed out that neither the integrity of the civil servants, nor of those engaged in the monitoring and verification process, should be taken for granted. Data and reports can be 'falsified' and, if large costs for companies are involved, illegal activities may arise.

A further critical element in monitoring, reporting and verification is the additional effort and cost necessitated by the high level of accuracy required in the data, which implies that trade-offs between monitoring costs and data accuracy will need to be struck. It may be thought preferable not to try to monitor each and every emission, but to settle for a high but imperfect level of accuracy and completeness. If monitoring costs vary significantly between various types of greenhouse gas emissions, the inclusion of a limited number of gases could be efficient in the face of overly high administrative and monitoring costs.

3. WORKING OF THE MARKET (THE CASE OF STATE-OWNED ENTERPRISES)

Emissions trading systems are based on the working of market forces. They are a cost-effective means of reducing abatement costs by allowing polluters to determine themselves who has the lowest abatement costs and who will take abatement action based on the respective price incentives. Price incentives, however, may not give rise to abatement by emitters that have the lowest abatement costs if decision makers base their conclusions on considerations other than emission allowance prices. Such behaviour could occur if the links between politicians and industry are too amicable. Examples of such close ties may be found in the Japanese context in descriptions of the Iron Triangle (the close relationships between business, bureaucrats and politicians that were much stronger several years ago) and state-owned enterprises (SOEs) in China.

In the case of SOEs, for example, it is conceivable that investment decisions on abatement technology will not be based exclusively on market considerations, but are in part motivated by policy considerations, which could include (regional) prosperity and the creation of (regional) employment. In such a context, one SOE might invest in abatement

---

[14] On this see Peeters (2006).

technology even though it might be more efficient for another firm to abate. Complying with the intensity targets under the Chinese pilot schemes is, after all, an important objective and if it might not be achieved via the market, it could be achieved by influencing the decision-making process of SOEs. Such command and control-like influence over the market price mechanism is likely to give rise to allocative inefficiencies and undermines the cost-effectiveness of the emissions trading scheme. Similarly, a SOE could be asked to buy emission allowances from another SOE, in order to support it financially, on terms it would normally not have accepted (a form of subsidy). Such influence will give rise to the same consequences in terms of negative efficiency.

Another critical issue that could arise in the context of state-owned enterprises, or regulated industries in general, is that these entities are unable to increase their prices in line with increases in production costs. Price increases are a natural response to the internalization of carbon costs and are dependent on the price elasticities of demand and supply. They can also be prevented by administrative guidance or outright price regulation. If, as a result of such instances, prices are not increased, the profitability of the affected enterprise is reduced. It will also need to invest more in abatement technology in order to mitigate the reduction in profits. Furthermore, consumers are not incentivized to refrain from environmentally hazardous activity.

Interference with the free working of the market price mechanism could therefore severely undermine the efficiency of the emissions trading system, which is something that should be considered by system designers and policy makers and should be avoided. It will be interesting to see how emission trading systems are able to function in economies in which other policy considerations play an important role in the entrepreneurial decision-making process.

4. TRANSACTION LOGS

Emission allowances exist in electronic form. Transactions between parties must therefore be recorded in an electronic database, commonly referred to as a 'registry'. A registry has the task of ensuring accurate accounting, and keeping track of the issuing of allowances and the operators' holdings, transfers, purchases, cancellation and submission of allowances.[15] Transaction logs are a vital element of any emissions

[15] This section is based on Jakob-Gallman (2011) p. 24.

trading system and provide the basis for tracking and monitoring transactions. Registries are therefore crucial for ensuring both the operation of an ETS and its environmental integrity. The latter can be impaired if surrendered allowances were to re-enter the market (double counting).[16]

At the international level there is the so called International Transaction Log (ITL).[17] This is operated by the UNFCCC and records transactions between the national registries of Kyoto Protocol Annex B countries and the CDM Registry. The CDM Registry is administered by the UNFCCC secretariat and holds newly issued certified emission reductions (CERs).[18] By means of the ITL, the UNFCCC is able to ensure the validity of transactions taking place within the framework of the Kyoto Protocol. Nations allowing for the use of CDMs must have a registry within the ITL and its national registry system must be able to work alongside the ITL.

Until recently the registries of the EU Member States were connected by the Community Independent Transaction Log (CITL). During the second trading period of the EU ETS (2008–12), which coincided with the Kyoto Protocol compliance period, the Member State registries functioned also as registries for the purpose of the ITL.[19] During the third trading phase (2013–20), the national registries of the Member States and the CITL are replaced by an EU-wide registry: the European Union Transaction Log (EUTL). Of course, the Member States of the EU will continue to be able to take part in the ITL, thus granting them access to CERs and ERUs.[20]

The EU ETS has introduced certain restrictions and given its EUAs certain properties that make them different from the normal assigned amount units (AAUs) under the Kyoto Protocol. During the Kyoto compliance period EUAs are created from AAUs from the EU Member States. The AAUs that are converted into EUAs therefore need to be tagged with additional information to distinguish them from ordinary AAUs; without such additional information this conversion could not take place and the EU ETS would not be able to deviate from the Kyoto

---

[16] The risk of 'double counting' occurred when Hungary legally sold surrendered Kyoto emission allowances to third countries. This issue is explained in Chapter 6.

[17] See Decision 13/CMP.1, Annex, para. 38.

[18] Ibid., Appendix D.

[19] See Jakob-Gallman (2011) p. 84.

[20] See Directive 2009/29 EC, Art. 19(1).

Protocol.[21] Emissions trading schemes seeking to establish restrictions on the quality of the allowances that can be used need to set up a registry system that is capable of administering such information.

One of the most important and obvious requirements of a transaction log is that it must be accurate and resistant to fraud and cybercrime. The accuracy of the data can be impaired if transactions have to move through several registries, or if the transaction log is interlinked with several other registries, as this will increase the likelihood of error. At the same time, connections with different transaction logs could increase the vulnerability of the registry to fraud or cybercrime affecting the system. After cyber attacks against national registries, trade in the EU ETS was suspended in the summer of 2011 until the national registries proved to the European Commission that they had adequate safeguards. When establishing or linking with any transaction log, operational safeguards and the prevention of fraud or cybercrime should be a leading concern for ETS designers.

In essence, transaction logs are electronic databases, which may contain very detailed information on transactions. To give a flavour of the kind of information that is contained in a transaction log, we present here some of the key elements of the CITL. At the start of the first trading period, the CITL contained information on the types of account holder of the acquiring and transferring accounts. Next, the CITL discerns Operator holding accounts related to the emitting installations. Person holding accounts were used by actors such as banks, brokers and NGOs that were participating in the EU ETS. For enterprises that own multiple installations it is convenient to make use of such accounts, as they enable them to have their (allocated) allowances administered in one place, facilitating their allowance trade activities. Holding accounts are used by the national registry for adding installations' allocations, subtracting their emissions and for the New Entrance Reserve.[22]

---

[21] Bazelmans (2008) p. 300.

[22] Over 2005 and 2006, the CITL records the limited use of two other types of national registry account: Retirement Accounts and Voluntary Cancellation Accounts. Yet, according to the CITL, there are several more types of account: Non-Kyoto Account Type, Excess Issuance Cancellation Account, Mandatory Cancellation Account, Net Source Cancellation Account, Non-compliance Cancellation Account, tCER Replacement Account for Expiry, tCER Replacement Account for Reversal in Storage, lCER Replacement Account for Non-submission of Certification Report, lCER Replacement Account for Reversal in Storage (Type 2), Voluntary Cancellation Account (Type 3).

In addition to the types of account holder, the account holder names of the acquiring and transferring accounts are also registered. In most cases the name of the company, installation or the company's EU ETS representative is used. It is possible that a (unique) account holder name is attached to more than one Operator and/or Person holding account, or more than one holding account. There are also identifier codes for the acquirer and transferring account. The CITL also contains information on the identifier codes of the acquiring and transferring accounts, the countries' registries involved, the timestamp and ID of each (spot market) transaction, the number and codes of allowances transferred. Originally all allowances under the EU ETS cap were numbered, from the first to the last tonnes of CO_2. A transaction was then recorded as a batch. To deduce the amount traded, one needed the starting number (for example, 100,000) and the end number (for example, 105,000, resulting in 5,000 allowances). The result was that disjointed sets of such (starting and ending) allowance numbers were traded. As it was rather unlikely that the batches would be amalgamated back again (that is, through the end number of one batch and the starting number of another), with more frequent trading the batch size could only decline. More smaller batches were thus needed to make up a transfer. This was a potential source of error. It also allowed interested parties to track the flow of allowances through the system. This was deemed to be undesirable not only from a privacy perspective but also in terms of market efficiency in the context of stolen allowances. The identification of the allowance numbers has now been abolished.

Initially, the name and contact details of the company's EU ETS representative were made available in the CITL. This again led to privacy concerns, leading the EU Commission to remove the representative's name and (some) contact details. However, for a transaction from the account of an installation (an Operator holding account), the name, address, post code, city, and country of the installation site is available upon request.

Since installations belong to particular industry sectors, they were also classified by and attributed with industry activity codes. In the first phase of the EU ETS these were: 'combustion, mineral oil refining, coke ovens, metal ore roasting, pig iron/steel production, production of cement, glass, ceramics, paper and pulp, as well as a residual opt-in category'.[23] Over the years, more industry sectors have been added to the EU ETS, resulting in more activity types: aircraft operators, production of adipic

---

[23] Zaklan (2013).

acid, ammonia, bulk chemicals, carbon black, hydrogen and synthesis gas, nitric acid, primary and secondary aluminium, soda ash and sodium bicarbonate, gypsum and plasterboard. Since installations are attributed one code only, this may not be enough to provide accurate information on the sources of greenhouse gases of the respective installations. These codes are also not suitable for benchmarking purposes, which necessitates higher levels of detail. This could be provided, for example, by industry classification codes (these industry codes, upon which the Phase III benchmarking allocations are partly based, are discussed below). In addition to the above, the registry contains information on the permit for participation in the EU ETS system; this is in the form of a permit identifier and the permit date and, in most cases, it is a number uniquely coupled with an installation.

Besides vulnerability to fraud and susceptibility to errors, there are a number of other transaction log design issues that need to be considered when establishing links with other transaction logs. When setting up an emissions trading system, rules concerning offsets and other emission allowances need to be formulated. In addition, information requirements will need to be defined for trades. Linking may be difficult if the information in one transaction log differs from the data contained in a system with which one would like to establish an interface. Data needs to be complete to ensure full compatibility between the transaction logs.

Transaction logs should use one common script. While this may not be easily attainable – since different languages use different characters – the efficacy of the computer system will require the data to be easily accessed and transferred between different registries. A compatibility issue was indeed present in respect of data from the CITL as a result of the different scripts of the languages in the EU Member States.

Another issue that may obstruct the working of emissions trading systems, as well as their linking, is the fact that the setting up of registries is tedious and time-consuming. They need to be fully operational and tested before they can be used. System failures would undermine the trust of market participants and in emissions trading in general. In the EU ETS several Member States did not succeed in setting up their registry systems in time (the Polish registry, for example, only became operational in 2006),[24] and were thus only able to commence operation after delays. This teaches us that establishing the interoperability of schemes may also be time-consuming. The interoperability

---

[24] The first transaction from the Polish registry was registered with the CITL on 10 July 2006.

between registries or transaction logs could also be impaired if registries do not contain all of the information that should be contained. This was the case with regard to the Austrian and Greek registries in the first EU ETS trading phase.[25]

Transaction logs should not only be designed taking into account the current need of the emissions trading system, but preferably also future needs. Future needs can be determined by changes to the ETS itself or necessitated by linking with other systems. One way to allow for changes in the system is to include industry classification codes in the electronic database. In this way, a later change in the allocation method towards free allocation based on benchmarking or to a more favourable treatment of trade-exposed sectors is facilitated (that is, to prevent carbon leakage). In the case of linking with other transaction logs, however, it should be mentioned that there are different industry classification systems (for example, NACE, NAICS, SIC) that are used in various countries. Unless the classifications that are used are indeed identical, attention will need to be devoted to establishing comparable categories for the respective sectors and sub-sectors.

As can be inferred from the above, the data contained in the transaction log is vitally important and ETS designers need to pay attention to the design of an efficient and robust system that is also capable of meeting future challenges. Two other important issues, however, have not yet been addressed. These are information security and privacy considerations, as well as research on allowance transactions.

4.1 Information Security and Privacy

Information on the use and exchange of emission allowances is commercially sensitive information since it allows competitors a better understanding of their competitor's strategic position, their cash flow and their production facilities. Businesses therefore have a legitimate interest in the protection of this information and in restricting its publication. In the European Union the publication of information contained in the CITL was delayed by five calendar years.

The identification of sellers and buyers of allowances is further complicated by the chosen structure of the database. Registering only the names of the account holders makes identification of ownership

---

25 See Zaklan (2013).

structures a laborious and imprecise process that only a few will be willing to undertake.[26]

The measures to protect data seem to adequately reduce the usefulness of any published information. Business information therefore appears to be effectively protected. While business concern for the protection of data is expected to be uniformly present in all emissions trading systems and thus finds appropriate legal expression in the transaction log rules, the legal level of privacy and data protection is likely to differ.

Natural persons are allowed to participate in the EU ETS and are therefore eligible to become account holders. They can invest in the emissions trading market just as they can in stocks or other commodities. Admission requirements are stringent, of course, but once admitted private individuals can start to trade in emission allowances. Their trading behaviour can be identified online. We believe that this could give rise to some concerns for ETS designers.[27]

Privacy is a human right. Article 12 of the Universal Declaration of Human Rights of 1948[28] reads: 'No one shall be subjected to arbitrary interference with his privacy, family, home or correspondence, nor to attacks upon his honour and reputation. Everyone has the right to the protection of the law against such interference or attacks'. The wording of this Article is identical to the wording of Article 17(1) and (2) of the International Covenant on Civil and Political Rights,[29] which has been ratified by most nations.

The application of the right to privacy may, of course, differ between different jurisdictions. In the United States, for example, the right to privacy is more restricted. In the EU, the protection of the right to privacy is considered to be a fundamental right. Article 7 of the Charter

---

26 Note, however, that the European Commission has updated the transaction log so that the account holder names now have to be identical to the (full) names of the companies themselves. This facilitates the identification of those transacting.

27 Before elaborating upon this issue, note that this section builds on research by the author that is currently ongoing. The points made regarding privacy and data protection are in need of further refinement but our current insights suggest that these issues should not be disregarded by ETS designers. It is also important to point out that we have not yet looked at issues such as 'data minimization' and 'consent', which can constitute potential solutions to the issues raised.

28 Universal Declaration of Human Rights, UNGA Resolution 217A (III), UN Doc A/810 (1948) 71.

29 International Covenant on Civil and Political Rights, adopted by UNGA Resolution 2200 (XXI) of 16 December 1966, entered into force 23 March 1976, 999 UNTS 171.

of Fundamental Rights of the European Union (the Charter) regulates the protection of private and family life, home and correspondence and Article 8 regulates the protection of personal data. While the provisions of the Charter address Member States only in so far as they implement Union law,[30] their membership of the European Convention on Human Rights (ECHR or the Convention)[31] extends the application of the protection of the right to other situations. Article 8 of the Convention guarantees protection of the private life of individuals from abusive interference by public authorities.

The provision of Article 8 of the Convention, however, does not refer explicitly to the right to privacy as such, but to four different, not mutually exclusive, areas of personal autonomy: private life, family life, home and correspondence.[32] Despite the large number of decisions based on this Article, the European Court of Human Rights (the Court) has not given a conclusive definition of the right. With regard to 'private life' the Court has stated in several judgments that it is neither necessary nor possible to give an exhaustive definition of it.[33] Private life is thus to be considered as a broad term, the content of which will be established on a case-by-case basis.[34]

Article 8 of the Convention makes no explicit separation between the right to privacy and the right to data protection, as is the case with the Charter of Fundamental Rights of the EU.[35] The Court has been implicitly making such a distinction in its decisions. For the Court, the mere collecting and storing of information relating to an individual's private life by a public authority amounts in itself to an interference with the right to privacy. It does not matter if the data has been subsequently used, or the way in which it was used.[36]

---

30 See Charter of Fundamental Rights of the European Union, Art. 51(1).

31 European Convention for the Protection of Human Rights and Fundamental Freedoms, entered into force 3 September 1953, 213 UNTS 222.

32 This section is based in part on Milaj (2013).

33 *Niemietz v Germany*, ECHR application no. 13710/88, 16 December 1992, para. 29; *Peck v United Kingdom*, ECHR application no. 44647/98, 28 January 2003, para. 57; *Pretty v United Kingdom*, ECHR application no. 2346/02, 29 April 2002, para. 61.

34 Ibid.

35 See Art. 7 (respect for private and family life) and Art. 8 (protection of personal data) of the Charter of Fundamental Rights of the European Union; De Hert and Gutwirth (2009).

36 *Leander v Sweden*, ECHR application no. 9248/81, 26 March 1987, para. 48; *Kopp v Switzerland*, ECHR application no. 23224/94, 25 March 1998, para.

The right to privacy as established by Article 8 ECHR is not an absolute right.[37] The second paragraph of the article lists the limits of its application. According to this provision, interference by public authorities with the private sphere of individuals is permissible if it is carried out in accordance with law, is necessary in a democratic society and is counterbalanced by one of the following interests: national security, public safety, the economic well-being of the country, prevention of disorder or crime, protection of health or morals, or the protection of rights and freedoms of others.

That the rules on the protection of privacy may be of relevance in the context of transaction logs can be seen from the European experience. Under the former CITL rules, it was possible to relate the identity of a private investor or trader with particular trading behaviour. This information can be linked to general market developments and hence be used as an indication of the relative success or failure of an individual, or reflect upon his particular trading or business strategy. Privacy and data protection were therefore issues and the European Data Protection Supervisor demanded amendments of the law.[38] Differences in data protection requirements should therefore be taken into account when establishing transaction logs.

It is not only privacy that may be impaired: rules on data protection can also present problems. In the European Union, for example, the draft General Data Protection Regulation[39] envisages strengthened restrictions on the movement of personal data to third countries or international organizations.[40] Data from the EUTL can flow to a foreign registry only if the necessary standards for the protection of data and privacy are met. This may cause problems, in particular, if data exchanges between Australia and the EU are extended to US emissions trading systems, such as California, where the level of protection for data and privacy is lower. It can thus be summarized that both privacy and data protection rules ought to play an important role in the context of transaction logs for natural persons, and consequently also for the linking of emissions trading systems.

53; *Amann v Switzerland*, ECHR application no. 27798/95, 16 February 2000, para. 68.

[37] Kleining et al. (2011) p. 43; Kilkelly (2001) p. 6; Himma (2007).

[38] EDPS Opinion (2012).

[39] General Data Protection Regulation, COM(2012) 11 final, 2012/0011 (COD).

[40] See Costa and Poullet (2012) p. 261.

4.2 Research on Allowance Transactions

The corollary of having a substantial degree of privacy protection is that the emissions trading market is less transparent. This means that less information is available to conduct research on the market and to point out to policy makers where potential problems might lurk, apart from the fact that for market supervision efforts detailed data is needed in order to ascertain market abuses.[41] In order to offer such information to policy makers, a number of transaction log design issues should be considered.

First, long delays in the publication of data reduce the policy relevance of any research conducted on the emissions trading system. If policy makers wish to benefit from research, they must either provide the data to research contractors and pay for the research or shorten the publication time lag of the data. In the case of the EU ETS, it may indeed be wondered whether a five-year publication delay allows policy makers to benefit from the research that is being conducted.[42] Under the new EUTL regulation this time delay has been reduced to three years,[43] but the same argument could be just as relevant for this time period.

Second, if data is renewed or revisited in the transaction log, this should be indicated. If changes in the data are not indicated, it will negatively affect the validity of the research findings.

Third, an important element in the use of transaction data for research concerns the uniqueness of transactions. Even though Operator accounts can be associated with specific installations, the ownership structure of these installations is more difficult to ascertain. In analysing transactions between installations, researchers may be unable to manipulate the data in such a way as to correctly associate all installations with their correct parent company. One possibility would be to include parent companies in the transaction log or to include the tax codes of the companies.[44]

Fourth, there is little transparency in actual transactions. For example, in order to prevent market abuse and market manipulation, effective

---

[41] If detailed data is collected, legal rules must be in place to grant authorities charged with the task of supervising the market full access to the data.

[42] CITL transactions were made available on 1 January in the fifth year after the recording of the information according to Annex XIII(4) of Commission Regulation (EU) No. 920/2010 and Annex XII(4) of Commission Regulation (EU) No. 1193/2011.

[43] Commission Regulation (EU) No. 389/2013 of 2 May 2013, Table XIV-I, point 4.

[44] Both were proposed during the CITL workshop of the 2012 Climate Policy Conference at the European University Institute. See http://fsr.eui.eu/Events/ENERGY/Workshop/2012/121003CITLDataWorkshop.aspx.

oversight structures need to be established to prevent any undue conduct. Such structures can only work effectively if they have relevant market data at their disposition. In the case of the EU ETS, neither the CITL nor its revised EUTL have such data recorded. Neither requires the recording of exact transaction prices. Also the type of the transaction (over-the-counter exchanges, brokerage or transactions at an exchange) is not recorded. Institutions charged with monitoring markets lack important information both on the ultimate identity of traders and on price, which severely impedes the execution of effective market oversight. Obviously it also severely impedes research by third parties on this topic. The appropriate information for conducting market analysis is thus not available.

As can be seen from the above, policy makers have the task of balancing the objectives of trade secrets and protection of privacy, on the one hand, with research that can inform policy making and support effective market oversight, on the other.

5. CONCLUSION

This chapter has presented several issues relating to the operability of an emissions trading system that designers should consider. With regard to MRV, it has been pointed out that designers will need to strike a balance between (very) high levels of data accuracy, on the one hand, and monitoring costs, on the other. Monitoring is essential for the operation of any ETS since it is the precondition for a level playing field between companies. Reliance on self-monitoring and self-reporting calls for more rather than less effective oversight. In addition, the integrity of civil servants or other people engaged in the monitoring and verification process is not to be taken for granted.

With regard to SOEs, it has been stressed that the benefits of emissions trading markets derive from abatement by those operators with the lowest abatement costs. If operators are subject to administrative guidance or otherwise incentivized not to follow their best entrepreneurial insights, the ETS will not be able to deliver cost-effective emissions abatement.

The final issue presented in this chapter is that of transaction logs. Transaction logs are a technical and highly complex area of climate law where the 'devil' will frequently be found in the detail, but this does not make them any less important. What is made abundantly clear in this treatment is that transaction log systems must be tested and fully operational before going online. Furthermore, a number of trade-offs that ETS designers will need to make have been identified: the level of transparency of an emissions trading market may be in conflict with privacy and data protection rules for natural persons, or may limit the input that policy makers may receive from market research institutions.

8. Implementation issue 4: Lawsuits following from emissions trading

1. INTRODUCTION

The preceding chapters have made clear that whenever a policy measure is taken there will be stakeholders who are pleased and some who are not. This chapter seeks to present some of the legal challenges that have arisen in the context of emissions trading design, in particular in relation to climate change. The reason why we have chosen to extend the chapter to cover climate change litigation rather than confine it to emission design issues is because emissions trading is simply an expression of a wider policy. Emissions trading will frequently be employed in conjunction with other measures. Challenges to one element of a nation's climate change policy may impact on other elements, including an ETS.

Climate change litigation is a fairly recent phenomenon. The first of such cases emerged in the United States in 1990,[1] and in recent years there has been an increase in the number of cases touching upon climate change issues. Even though such claims are now more widely accepted by the courts, there are still cases where the underlying scientific evidence is questioned. It is noticeable that the cases in which climate change is touched upon become more and more diverse.

Given the diversity of the issues and the proliferation of existing and emerging schemes, it is not possible to present a complete overview of all legal issues in one brief chapter. Our ambition is therefore more modest: we seek to give a flavour of the types of legal challenge that can arise in the context of emissions trading design. This chapter shows that a wide variety of issues have been challenged and that the legal dimension of designing an emissions trading system may be easily underestimated.

The case law presented here comes from various jurisdictions and therefore relates to different emissions trading systems. Naturally the outcome of these cases depends on the particular legal system. The aim

[1] *City of Los Angeles v National Highway Traffic Safety Administration*, 912 F 3d 478 (DC Cir., 1990).

here is not to elaborate upon the legal systems themselves but to present potential legal claims so that they can make policy makers aware of the importance of law when designing emissions trading systems, and potential problems. The outcome of each case is therefore less relevant here than the actual problem that is identified. Naturally this chapter is not directed towards lawyers or judges, but towards emissions trading designers.

It is also important to stress that the motivation for starting a lawsuit can vary. The mere fact that a case on a particular issue has been started does not imply that there is in fact a legal problem. It may constitute strategic behaviour on the part of a plaintiff who, for example, seeks to obstruct the implementation of an undesired policy.

Given the huge diversity of the legal issues and actors involved, we decided not to present the cases on the basis of the emissions trading systems and jurisdictions in which they have arisen, but on the basis of the issues we have identified. First, we consider the 'object' of the legal claim. Second, we highlight specific legal areas that have given rise to specific types of legal issue. Third, we consider critical issues that are capable of shaping legislative choices when designing an emissions trading system.

In Section 2 we review, among other things, statutory challenges, challenges against the competences of the enforcement agencies and challenges against the implementation of the law. Section 3 focuses on legal issues such as liability, property rights and consumer protection. Finally, in Section 4 we highlight particular critical issues that may shape the design features of emissions trading systems as they impact on judicial review – namely legal standing, legitimate expectations and policy objectives related to climate change.

2. THE OBJECT OF THE LEGAL CLAIM

The climate law cases presented in this section are structured according to their legal object. We consider statutory challenges, challenges against the competences of enforcement agencies, challenges against the implementation of the law, and challenges on the grounds of international law. We also consider challenges undertaken by the enforcement agency itself.

2.1 Statutory Challenges

Statutory challenges can have many directions. In the EU context they can address the validity of the law itself or the legality of particular elements that it regulates. This was at issue in the *Arcelor* case.[2] Arcelor objected that the steel sector was covered by the EU ETS Directive and the subsequent French implementation measures, while the aluminium and plastics industries were not. Arcelor alleged, inter alia, an infringement of the principle of equal treatment enshrined in the French Constitution. Within the framework of a preliminary ruling question, the Court of Justice of the European Union (CJEU) was asked to determine if the Community legislator had violated the equal treatment principle by extending the EU ETS to steel but not to the aluminium and plastics industries. The Court held that the sectors were in a comparable position while being treated differently. Given that non-covered sectors were not subject to comparable international or Community measures, this gave rise to a disadvantage for covered sectors that could not be offset by purely national measures. The Court examined whether the unequal treatment was justified on the basis of an objective and reasonable criterion and if it was proportionate to the aim pursued. This was the case for both aluminium and plastics. The exclusion of the chemical sectors was justified, given that the size of the industry would have made the administrative burden of including it in the EU ETS too large.

Opportunities to challenge statutes can sometimes come from more unexpected directions. Instead of directly challenging the statute itself, the science upon which the law is based may be challenged. The New Zealand Climate Science Education Trust filed a case against the National Institute of Water and Atmospheric Research Limited (NIWA) – a government-owned company which has maintained a national climate database since 1992.[3] The Trust challenged the validity of the data that was used as a basis to formulate laws. For climate change data to be subject to judicial review, the plaintiff needed to demonstrate 'some defect in NIWA's decision-making process or show that the decision was clearly wrong in principle or in law'. In this case, the Court proceeded to examine the credentials of the plaintiff's expert witnesses and concluded that they were not climate change experts. This case is interesting since it

---

[2] C-127/07 *Arcelor Atlantique and Lorraine and Others v Premier Ministre, Ministre de l'Écologie et du Développement Durable, Ministre de l'Économie, des Finances et de l'Industrie* [2008] ECR I-9895.

[3] *New Zealand Climate Science Education Trust v National Institute of Water and Atmospheric Research Ltd* [2012] NZHC 2297.

suggests that climate change science, as the basis of legal activism, is challengeable. If the science is wrong, policies and laws based upon it may be politically hard to maintain.

Other cases do not so much challenge the statute itself but rather the competence of the legislator to legislate, or claim that the legislator has infringed other laws by the legislation.

Two cases that address the competence of the legislator are *Thrun et al.* and *Indeck Corinth*. In 2011, *Thrun et al.*[4] challenged the legality of New York's participation in the Regional Greenhouse Gas Initiative (RGGI). The plaintiffs argued that New York's RGGI programme was *ultra vires* because the Memorandum was executed and the regulations were promulgated by the New York Department of Environmental Conservation (DEC) and New York State Energy Research and Development Agency (NYSERDA) without the consent or authorization of the legislature. Furthermore, the plaintiffs contended that the RGGI imposed a tax that was not authorized by the legislator and that the RGGI memorandum violated the Compact Clause of the US Constitution,[5] which states that '[n]o State shall, without the consent of Congress … enter into any Agreement or Compact with another State'. Because of the plaintiffs' lack of standing, the Court did not address the substantive issues. Also, in light of the unreasonable delay in pursuing the claim, the plaintiffs might have been time-barred (the so-called 'doctrine of laches'). *Indeck Corinth*,[6] a gas-fired combined cycle co-generation facility located in Corinth, claimed (among other issues) that entering into the Memorandum of Understanding to establish New York's participation in the RGGI was unconstitutional and *ultra vires*. The case was settled by the parties without resolving the question of constitutionality.

In addition to direct challenges to the competence of the legislator, it has been argued in some cases that the legislator did not comply with hierarchically superior rules of law. This was the case in the Dutch NOx case. The Netherlands established a NOx performance standard rate (PSR) emissions trading system to address acid rain. Under this system a company earns emission credits if it over-complies with the emissions standard, or needs to purchase additional credits if it falls short of it. The government takes an active role in monitoring but does not participate as an actor. The European Commission decided that the NOx system constituted State aid. This was challenged by the Netherlands. The Court

---

4 *Thrun et al. v Cuoma et al.*, Index No. 4358-11; RJI No. 01-11-104776.

5 Art. 1, para. 10, clause 3.

6 *Indeck Corinth, L.P. v David A. Paterson et al.*, Consent Decree, Index No. 5280-09.

of First Instance (CFI) overturned the Commission's decision on the grounds of selectivity and the Commission appealed. The European Court of Justice (ECJ) overturned the CFI's decision and concluded that the PSR system is selective and does indeed involve State aid. Interestingly the Court held that the Member State should have opted for a different measure – emissions trading with allowance sale or auctioning rather than a PSR system – in order to avoid the granting of aid. The Court thereby brought elements of the justification test under Article 107(3) of the Treaty on the Functioning of the European Union (TFEU) into the analysis of Article 107(1) TFEU. This suggests that Member States do not enjoy the discretion to select a policy measure that does not raise or even maximize profits if a different measure could be designed that would. Hence the discretion of EU Member States to design appropriate policy instruments appears to be restricted by this interpretation of State aid rules.[7]

A federal legal system is characterized by certain divisions of power and by a strong hierarchy of laws: federal law superseding state law. Nevertheless, regulation by different actors that in principle enjoy different competences can lead to overlapping and conflicting regulation in situations, for example, where climate policy touches upon other environmental policies. *Green Mountain Chrysler Plymouth Dodge Jeep v Crombie*[8] was the first US case to rule on an alleged conflict between federal fuel economy laws and greenhouse gas emission standards for new motor vehicles. Stakeholders from the automotive industry challenged Vermont's adoption of California's Greenhouse Gas standards on several grounds. The plaintiffs argued, in particular, that Vermont's actions (i) contravened the Clean Air Act, (ii) contravened the Environmental Protection and Conservation Act, and that (iii) Vermont was not allowed to enact greenhouse gas regulation since it constituted an internationally negotiated issue. Even though the automobile industry dropped the case at the appeal stage because of advancements in federal legislation, the case is interesting since it highlights that incongruities between state level and federal legislation can trigger serious legal challenges. Furthermore, it bears mentioning that in this case the Court followed *Massachusetts v Environmental Protection Agency*[9] and rejected the plaintiffs' argument that states were barred from enacting greenhouse gas legislation on grounds of foreign policy.

---

7 For an analysis of this issue see Weishaar (2013).

8 *Green Mountain Chrysler Plymouth Dodge Jeep v Crombie*, F. Supp. 2d 295 (12 September 2007).

9 *Massachusetts v Environmental Protection Agency*, 549 US 497 (2007).

2.2 Challenges against the Implementation of the Law

Climate policies, and in particular emissions trading systems, are implemented by designated agencies. It lies within the nature of their tasks that institutions will need to take actions and decisions. These are frequently subject to lawsuits claiming that the laws on which the policies are enacted must be observed or that alleged mistakes have been made. Other challenges relate to the competences and discretion of the institution or the policy design choice it has taken. Access to information also constitutes an important ground on which to challenge enforcement agencies.

Challenges for non-observance of the law

Latvia challenged the European Commission decision to demand an amendment of its National Allocation Plan (NAPs) for the period 2008–12.[10] The question arose as to whether the Commission was time-barred from issuing its decision. Since each notification triggers a three-month time period for the Commission assessment and the Commission was one day late in requesting additional information on the Latvian NAP, the Court established that the Commission was indeed time-barred. As a consequence, the Latvian NAP was definitive and the Commission decision was annulled.[11]

A challenge alleging a mistake by the European Commission in the context of the harmonized allocation format of the third trading phase was made in 2011. The introduction of the more harmonized allocation format stopped the frequent challenges against the Commission Decisions on NAPs (discussed below), but as the case shows this change was not entirely problem-free. In *Hüttenwerke Krupp Mannesmann and Others v Commission*[12] the plaintiff challenged and sought annulment of the Commission's Decision determining the transitional Union-wide rules for harmonized free allocation for emission allowances. Regarding sintered ore, the plaintiffs alleged that the criteria for setting benchmarks had not been observed and that data from plants not belonging to the sector had been used to calculate a sectorial benchmark. They alleged, furthermore, that the benchmarks for hot metal breaches were inadequate. The Commission had failed to take into account the full carbon content

---

[10] T-369/07 *Latvia v Commission* [2011] ECR II-01039.

[11] The Commission appealed. Case C-267/11 *Commission v Latvia* is still pending.

[12] T-379/11 *Hüttenwerke Krupp Mannesmann and Others v Commission* (case still pending).

of residual gases resulting from iron and steel production in respect of their use for electricity generation. The plaintiffs argued that the Commission's reduction of the carbon content violated the EU ETS Directive. The plaintiffs advanced additional pleas in law, alleging breaches on the basis of failure to state reasons under Article 296 TFEU and the principles of proportionality and equal treatment. The case is still pending at the time of writing this book.

Competences of the institution

Several cases challenge European Commission Decisions regarding National Allocation Plans (NAPs) of both the first and the second allocation phases of the EU ETS. They touch upon the division of competences of the Commission's right to review the compatibility of the NAPs with the principles contained in Annex III and the discretion of Member States to implement the EU ETS Directive 2003/87/EC.

In *United Kingdom v Commission*, the European Commission rejected the UK's amendment of its provisional NAP on the basis that the Member State had not addressed the grounds for the incompatibility of the NAP with the legislation identified by the Commission earlier. The Court clarified that the Commission was not entitled to restrict a Member State's right to amend a NAP:[13] the 'double public consultation system' would otherwise be deprived of its effectiveness.

In other cases, the competence of the European Commission to alter the number of allowances allocated by a Member State has been questioned. This was at issue in *Poland v Commission*. The Court held that the Commission did not have the right to impose an emissions ceiling for a NAP.[14] The Commission relied upon its own facts and calculations to declare and to rely upon its own data, rather than on data submitted by the Member State. The Court held that the Commission's power for review under Directive 2003/87/EC was severely restricted and substantively limited to verifying the conformity of the measures taken by the Member State with the criteria set out in the law. The Commission did not have the right to replace data contained in a NAP with its own data.[15]

---

[13] T-178/05 *United Kingdom v Commission* [2005] ECR II-04807, para. 61.

[14] T-183/07 *Poland v Commission* [2009] ECR II-03395, para. 131. The appeal case C-504/09P *Commission v Poland* (nyr) was unsuccessful.

[15] T-183/07 *Poland v Commission* [2009] ECR II-03395, para. 120.

It is not only in the European context that the competences of enforcement agencies are subject to litigation. *Indeck Corinth*,[16] for example, challenged the Budget Trading Program (6NYCRR Part 242) and the CO_2 Allowance Auction Program (21 NYCRR Part 507). These are two sets of state-wide rules formulated by the Department of Environmental Conservation and the New York State Energy Research Development Authority to implement the RGGI. The laws were allegedly *ultra vires*, arbitrary and capricious, and not supported by a proper record. In addition, Indeck argued that it was suffering from financial hardship because it could not pass on its costs to consumers. The case was settled by the parties.

A specific subject of litigation in respect of enforcement agencies concerns challenges against specific design choices of climate policy. The California Air Resources Board is charged with the implementation of the cap-and-trade emissions trading system. In *California Chamber of Commerce and Larry Dicke v Air Resources Board*[17] the Board is being challenged as not having express or implied authority under the legislation to employ an auction mechanism. The plaintiffs describe this auction mechanism as having a 'revenue raising component' and as an 'unconstitutional tax'. The plaintiffs argue that the law only permits the Board to charge a limited fee to cover administrative costs and that the Board does not have sufficient power to raise revenue. At the time of writing the case is still pending.

Discretion of the institution

Agencies charged with the establishment or operation of an emissions trading system frequently enjoy a margin of discretion in fulfilling their duties. This margin of discretion is challengeable and has given rise to litigation.

The Association of Irritated Residents claimed that the California Air Resources Board did not duly take into account the costs and benefits of a cap-and-trade scheme and failed to consider regulatory alternatives. It was alleged that the Board thereby contravened the Global Warming Solutions Act and the California Environmental Quality Act. The Court held that the Board enjoyed a wider margin of discretion under the former, while the latter law provides for a procedure for an Environmental Impact Review under which a claimant can show abuse of

---

[16] *Indeck Corinth, L.P. v David A. Paterson et al.*, Consent Decree, Index No. 5280-09.

[17] *California Chamber of Commerce and Larry Dicke v Air Resources Board* (nyr) Case No. 34-2012-80001313.

discretion. Given the brevity with which alternative instruments were addressed, the Court granted an injunction to prevent the Board from undertaking activities regarding the cap-and-trade scheme until it had provided an analysis of the alternatives. The First Court of Appeals overruled the injunction, thus enabling the Board to finalize the regulation. Subsequently, the Environmental Impact Review was held not to contravene the laws.

Access to information

Other cases that have arisen in the context of climate change and emissions trading are in respect of the granting of access to information. Clearly, in order to enable private parties to prepare a case and to challenge regulatory agencies or statutes, access to information is necessary.

Diehm v Greater Taree City Council[18] is a case from New South Wales (Australia). It concerned the refusal to provide information regarding the proposed emissions trading scheme. The information was denied on grounds of public interest. Diehm requested information on the costs of the proposed scheme that were included in the domestic wage charge. The Administrative Court ordered the granting of access to information.

Not all information held by public authorities should be revealed. In *Lagerkvist v New Jersey Department of Environmental Protection*[19] a journalist sought access to information, including the 'results of the RGGI first eight carbon auctions and the names of all bidders, the price of each bid and the quantity and type of allowances sold to each successful bidder'. The Court established that the requested information fell within the ambit of the Open Public Records Act. It was, however, not convinced by the plaintiff's arguments that disclosure of this information would improve the auction process or promote transparency. As a result, the Court denied the request for disclosure of information as falling under the derogation of preventing disclosure of information which would give an advantage to competitors or bidders.

Access to information charges have been brought in the European Union. The Commission denied Saint Gobain Glass Deutschland access to information that it had received by the Federal Environmental Agency of Germany on the applicant's installations. Germany was required to

---

[18] *Diehm v Greater Taree City Council* [2010] NSWADT 241; 2010 WL 4111270.

[19] *Lagerkvist v New Jersey Department of Environmental Protection*, Docket No. MER-L-2185-10.

transfer information on the basis of the Commission Decision determining transitional Union-wide rules for harmonized free allocation of emission allowances. Saint-Gobain sued[20] the Commission for alleged violations of various provisions of Regulation 1049/2001 regarding public access and Regulation 1367/2006 on the application to Community institutions and bodies of the provisions of the Aarhus Convention on (a) access to information, (b) public participation in decision-making, and (c) access to justice in environmental matters, as well as (d) Article 296 TFEU (the duty to state reasons). At the time of writing the case is still pending.

2.3 Challenges against the Enforcement of the Law

Besides statutory challenges and challenges brought against enforcement agencies, policy makers can also be subject to litigation. Policy makers often need to balance different societal interests, such as the societal needs for energy supply and ecologically sustainable development. In *Haughton v Minister for Planning*[21] the Land and Environment Court in New South Wales (Australia) had to examine, inter alia, if the Minister's granting of conditional concept plan approvals for power plant projects were arbitrary, illogical and unreasonable, and that no reasonable decision maker would have made this decision based on the material available to the Minister. The Court stated that the *Wednesbury* test[22] for judicial review of a decision by a public authority was stringent and that the applicant had failed to demonstrate the required level of proof.

In another case from New South Wales, *Gray v Minister for Planning*,[23] the granting of planning permission was challenged on the ground that the Director General of the Department of Planning had failed to take into account the precautionary principle when accepting a mining company's environmental impact assessment. The Court held that the Director General had failed to take the Ecological Sustainable Development principles into account, in particular the principles of

---

20 Case T-476/12 *Saint Gobain Glass Deutschland v Commission* (case still pending).

21 *Haughton v Minister for Planning and Macquarie Generation*; *Haughton v Minister for Planning and TRUenergy Pty Ltd* [2011] NSWLEC 217.

22 *Associated Provincial Picture Houses Ltd v Wednesbury Corporation* [1948] 1 KB 223.

23 *Gray v Minister for Planning* [2006] NSWLEC 720; 152 LGERA 258; 2006 WL 3438287; [2007] ALMD 5959.

intergenerational equity and the precautionary principle. Consequently the decision was declared null and void.

Climate change policies are closely associated with the transition of energy production and the energy intensity of the economy. Energy production methods that emit less CO_2 are incentivized. Renewable energy production is, however, frequently subject to challenge from local residents. Residents seem to be afraid of (potentially) hazardous effects of carbon capture and storage (CCS), or are disturbed by the sight of large wind parks, or seek to avoid adverse environmental impacts on flora and fauna. While some of these claims are successful, others are not, as is demonstrated by the Australian case of *Taralga Landscape Guardians Inc v Minister for Planning.*[24]

In another case, the decision of the California Air Resources Board to use carbon offset credits is being challenged.[25] Both the offset provisions of the Board's cap-and-trade regulation and the individual offset protocols adopted to implement those provisions are being challenged as being incompatible with the California Global Warming Solutions Act. The plaintiffs contend that, by the use of 'market-based compliance mechanisms' to meet the requirements of the statute, the state legislator clearly did not intend that those mechanisms would include offsets from projects that achieve emissions reductions that are not in addition to reductions otherwise required by law or regulation, or that otherwise would occur in any event. As can be seen from this case, additionality is clearly one of the principal reasons for challenging the use of offsets for compliance purposes in an emissions trading system that seeks to address global warming. At the time of writing the case is still pending.

Challenges relating to particular design features of the EU ETS have arisen. For example, Member States have attempted repeatedly to include *ex post* adjustment provisions into their National Allocation Plans (NAPs), while the European Commission has successfully prevented such adjustments in the NAPs of several Member States. The German NAP was rejected in so far as it contained provisions allowing for *ex post* adjustments. In the resulting case, *Germany v Commission*,[26] the Court held that that the Commission had erred in law when it decided that the German *ex post* adjustments violated two criteria contained in Annex III

---

[24] *Taralga Landscape Guardians Inc. v Minister for Planning and RES Southern Cross Pty Ltd* [2007] NSWLEC 59; LGERA 1 2007 WL 441697.

[25] *Citizens Climate Lobby and Our Children's Earth Foundation v California Air Resources Board* (still at filing stage), San Francisco Sup. Ct., filed 27 March 2012.

[26] Case T-374/04 *Germany v European Commission* [2007] ECR II-4431.

of the EU ETS Directive. The criteria in question related to non-discrimination between companies and sectors (Criterion 5) and the requirement that the NAPs should contain a list of covered installations with the quantities of allowances intended to be allocated to each (Criterion 10); the Commission had incorrectly considered these to have been infringed. Despite the Court's ruling, the relevance of *ex post* adjustments in the first and second trading phases of the EU ETS was marginal. Regarding the case at hand, the impact on Germany in the second trading phase was zero since the relevant national laws for allocation under the second trading phase had already been completed. During the third trading phase, allocation is based on EU-wide harmonized allocation rules, where *ex post* adjustments – at least regarding closure and transfer rules – are shaped by the European Commission.

2.4 International Law Issues

Not only national or supranational law is relied upon to litigate in climate law contexts, but also international law. The extension of the EU ETS to the aviation sector and its associated extraterritorial application has led to strong international resentment (felt, in particular, by the United States and China[27]) and to a challenge before the English High Court for a preliminary ruling. The Air Transport Association of America and other parties have instituted proceedings in the English courts against the UK Secretary of State for Energy and Climate Change over measures implementing the EU ETS in the United Kingdom. In the course of these proceedings, a preliminary ruling question was submitted to the CJEU.[28]

The Court was asked if the extension of the EU ETS to include the aviation sector was in violation of international treaty law and customary international law. The Air Transport Association of America (hereinafter 'the Association') argued that such a scheme violated (i) the multilateral International Civil Aviation Convention (the 'Chicago Convention'), (ii) the bilateral US–EU Air Transport Agreement (the 'Open-Skies Agreement), and (iii) the Kyoto Protocol.

The Court found that the amended EU ETS Directive acquitted itself of alleged violations of international law. In relation to the Chicago Convention, the Court held that, although the Member States of the EU are

[27] The US passed a bill to prevent its airline companies from participating in the EU ETS: see European Union Emissions Trading Scheme Prohibition Act of 2011.

[28] Case C-366/10 *Air Transport Association of America and Others v Secretary of State for Energy and Climate Change* (2011 nyr).

signatories to the same, the EU itself is not a party and is therefore not bound by it.

With regard to the Open Skies Agreement, the Court found that, although the EU was bound by it, the provisions of the Directive did not violate Article 7 of the Agreement, which provides that EU law applies to aircraft when they enter the territory of a Member State. The Association argued that, as the Directive requires calculation of emissions based on fuel consumed during the entire flight, it covers the entire duration of an American flight, even when it passes over third states and the open seas; hence the Directive imposed an extraterritorial obligation. The Court reasoned that the Directive does not apply to aircraft registered in third states that fly only over third states and the high seas, and even over Member States, without stopping. Thus, it is only when a flight is registered in a Member State, or when an internationally registered flight arrives at or departs from a Member State, that it falls within the purview of the Directive. The Association also argued that the Open Skies Agreement provides exemptions from duties and other charges on fuel loads; the Court dismissed this argument, observing that the EU ETS is a market-based environmental measure and not a duty or a charge on fuel loads. The Court also observed that the Open Skies Agreement allows for environmental reasons to exclude its applicability, but this observation was not relied on in the judgment.

Article 2(2) of the Kyoto Protocol requires the limitation or reduction of emissions of certain greenhouse gases from aviation bunker fuels, working through the International Civil Aviation Organization (ICAO). The Court held that the Kyoto Protocol imposes quantified commitments: '[t]he parties to the Protocol may comply with their obligations in the manner and at the speed upon which they agree'. Thus, for all practical purposes, the adoption of the provisions of the Kyoto Protocol, other than those relating to the targets, are left to review by the EU; with regard to Article 2(2), it was found to be not 'unconditional and sufficiently precise so as to confer on individuals the right to rely on it in legal proceedings'.

The two principles of customary international law that were invoked were: (i) no state may subject the high seas to its sovereignty, and the corollary freedom of states to fly over the high sea, and (ii) each state has complete and exclusive sovereignty over its airspace. The Court held that it is a principle which applies to the high seas (and flagships), and not necessarily to the airspace above the high seas. The Court also observed that the United Kingdom and Germany do not agree that this is a principle of customary international law. Generally speaking, the Court held that since a principle of customary international law does not have the same degree of precision as a provision of an international agreement, judicial review must be limited to considering whether the institutions of the EU made manifest errors of assessment concerning the conditions for applying those principles. In this regard, the Court reiterated its position that the Directive applies only to flights 'which arrive at or depart from aerodromes situated in the territory of the Member States', and thus does not breach the sovereignty of third states and freedom to fly over the high seas.

Despite this ruling, international resentment has continued. To improve the relationship with other countries and to provide a 'positive atmosphere'[29] for negotiations, the European Commission proposed a suspension of the compliance duty on flights into and out of the EU for one year[30] until the ICAO General Assembly in September 2013.[31]

Other issues may crop up under international law. Cases may arise at the interface between an emissions trading system and the rules of the World Trade Organization (WTO). Of particular interest here will be the General Agreement on Tariffs and Trade (GATT), the General Agreement on Trade in Services (GATS) and the Subsidies and Countervailing Measures (SCM) Agreement. It might be questioned whether grandfathering emission allowances for free could constitute a subsidy within the WTO framework. Countries may be barred from introducing carbon cost-related border tax adjustments to compensate for the lack of stringent environmental standards abroad.

2.5 Challenges Brought by Enforcement Agencies

The European Commission acts as the 'guardian' of the EU Treaty. Under Article 258 TFEU, the Commission is empowered to bring cases concerning the failure of Member States to correctly implement a Directive. Within the context of the EU ETS, the European Commission has taken action against Italy and Finland. Italy failed to adopt in time all the laws, regulations and administrative provisions necessary to comply with the Directive.[32] Finland implemented the necessary laws in time, but provided exemptions for the province of Åland. The Court ruled that Finland had failed to comply with its obligations under the Treaty.[33] Most

---

29 European Commission (2012b).
30 COM(2012) *697*.
31 European Commission (2012c).
32 C-122/05 *Commission v Italy* [2006] ECR I-00065.
33 C-107/05 *Commission v Finland* [2006] ECR I-00010.

cases of non-implementation of European law, however, are resolved by the parties without ever reaching the CJEU.

3. SPECIFIC LEGAL AREAS

This section presents particular areas of law that deserve attention from the policy makers engaged in climate change decisions. The legal issues are liability, property rights and consumer protection.

3.1 Liability

Liability issues are frequently raised within the framework of tort-based claims against emitters of greenhouse gases. Comer and several other individuals brought an action against electricity utilities, and coal, chemical and oil companies for damages for loss resulting from Hurricane Katrina. The action was based on public nuisance, negligence, unjust enrichment, civil conspiracy, fraudulent misrepresentation and concealment, and trespass. The public nuisance claim was based on the defendants' alleged contributions to global warming and hence to the ferocity of the hurricane.

The case was dismissed for lack of standing since the plaintiffs could not prove that their injuries were 'fairly traceable' to the defendants' activities and on the basis of non-justiciability of political questions.[34] The Appeal Court overturned the judgment by granting standing,[35] but the Court lost its quorum and procedural rules required the case to be dismissed. The lower court's decision therefore remained. In 2011 the case was brought again before the lower court and dismissed on the basis of the doctrines of *res judicata* and collateral estoppel. The Court cautiously re-examined the issue of standing and affirmed its 2007 decision that the plaintiffs did not have standing because they could not prove that their injuries were fairly traceable to the defendants' activities and affirmed that the plaintiffs were barred by the political question doctrine. It was also held that the state law on which the claims were based was pre-empted by the Clean Air Act.

In *Kivalina v Exxon Mobil Corporation et al.*,[36] the residents of the Alaskan Native Village of Kivalina filed a lawsuit based on nuisance

---

34 *Comer v Murphy Oil USA*, Case No. 1:05-CV-436-LG-RHW.

35 *Comer v Murphy Oil USA*, WL 2231493 C.A.5 Miss.

36 *Native Village of Kivalina v Exxon Mobil Corporation et al.*, 2009 WL 3326113.

against American oil and electricity utility companies for destruction of their village by flooding caused by climate change. The Court dismissed the case on the ground that regulating greenhouse gas emissions is non-justiciable: this is an issue to be resolved by Congress and the Administration rather than by courts. It bears mentioning that the Court found the plaintiffs' arguments unconvincing – that is, that the defendants contributed to climate change (contributory causation) and the flooding. In the reasoning of the Court, greenhouse gases were undifferentiated and could not be attributable to a particular defendant. An appeal was filed with the Ninth Circuit Court of Appeals in November 2009. In September 2012, the panel of appeal judges decided not to reinstate the case.

The prospects for the success of liability claims in respect of climate change damage in the US are poor. Cases are barred on the grounds of non-justiciability of political questions, and they encounter problems of standing, causation, attribution and retroactivity. Liability claims are not restricted to private parties: governments and state entities can also be subject to litigation. Holcim (Romania) SA, for example, has sued the European Commission for the market value of and the interest in stolen emissions that would not be recovered by the time of the judgment. The plaintiff alleges that the Commission took an unlawful decision not to disclose or allow the disclosure of the location of stolen emission allowances in the framework of the EU ETS.[37]

In a more general vein, it may be asked if public authorities could be held liable for not adequately addressing climate change, or if private companies could be saved from liability by reference to their compliance with legislative standards or participation in an emissions trading system. Different jurisdictions might reach different solutions for such legal questions.

3.2 Property Rights

Emission allowances or permits are integral elements of any emissions trading system. Even though allowance trading systems have been operating for several years and various forms of emission rights have emerged, their property rights are still not properly understood. They are generally regarded as intangible property rights (at least on the basis of CJEU case law). What remains unclear is the exact nature of the property rights they contain and how the rights differ among the various emissions

---

37 T-317/12 *Holcim (Romania) SA v European Commission* (case still pending).

trading schemes and across the Member States of the EU. The latter, in particular, is interesting in the context of the IT fraud (phishing) attacks on account holders that took place in January 2010 and the theft of allowances from national registries of several EU Member States that took place in January 2011. Can account holders acquire property rights in stolen emission rights? If so, under what circumstances? Are there varying levels of legal protection of property in the jurisdictions of the Member States regarding purchases of emission allowances?

Armstrong DLW GmbH, a German subsidiary of an international flooring manufacturer, sued Winnington Networks Ltd for propriety restitution, unjust enrichment and unconscionable receipt of trust property.[38] The defendant had bought 21,000 EUAs from Zen Holdings Ltd, which had posed as the legitimate owner of the allowances. Although Winnington had requested Zen to provide information about the company and title to the allowances, the defendant sold them before receiving such information.

The case was adjudicated on the basis of equity where even if a thief acquires legal title, the equitable title remains with the victim. The Court rejected the defendant's argument that the company had acted *bona fide*, holding that the company 'had deliberately closed [its] eyes' to the possibility that the allowances were stolen and had 'willfully and recklessly failed to make the enquiries'.

3.3 Consumer Protection

Accompanying the creation of an emissions trading system is the heightened receptiveness of consumers to purchase 'green' and 'carbon-friendly or -neutral' products. Companies even offer consumers the possibility to offset their own flight or car travel. Consumer law claims could then arise in relation to climate mitigation promises.

One such case concerns GM Holden, a supplier and dealer of Saab motor vehicles in Australia.[39] Holden advertised that planting 17 trees for every Saab vehicle purchased would make the vehicle carbon-neutral. The Australian Competition and Consumer Commission (ACCC) sued Holden for breach of the Trade Practices Act by false advertisement that indicated that Saab had novel features that made it climate-neutral, and

---

38 *Armstrong DLW GmbH v Winnington Networks Ltd* [2012] EWHC 10 (Ch).

39 *Australian Competition and Consumer Commission v GM Holden Ltd* [2008] FCA 1428; 2008 WL 4261039.

alleged that 17 trees would not compensate for a vehicle's emissions for more than a year.

Prior to any pronouncements by the Federal Court on whether it could review Holden's offset policy, Holden informed the ACCC that it would withdraw the advertisement and would plant 12,500 trees to offset the carbon emissions for the life of all Saab motor vehicles sold during the advertising campaign.

4. CRITICAL LEGAL ISSUES IN ENACTING AN ETS

This section examines a number of critical issues that policy makers should bear in mind when considering enacting an emissions trading system. The issues relate to legal standing before the courts and the legitimate expectations of interested parties, and hence with the opposition that any climate change action could face. The precautionary principle, which should guide good governance and policy making, is also discussed.

4.1 Legal Standing

Laws that establish emissions trading systems belong to the realm of public law. It is therefore not surprising that challenges against these laws are typically subject to restrictive requirements for legal standing before the courts. Legal standing requirements differ between jurisdictions, and also between laws. The US Clean Air Act requires stationary sources to achieve air pollution standards necessary for the protection of public health and welfare, and requires the Environmental Protection Agency (EPA) to develop and enforce regulations to protect the public. The EPA, as well as any citizen or government body or institution, may file a lawsuit under the Clean Air Act to enforce compliance or sanction any omission. Impediments to legal standing are therefore relatively low and many cases have been filed under this Act to challenge the activities of the EPA or its regulations.

By contrast, the requirements for legal standing under the US National Environmental Policy Act seem to be higher. This is a federal law that has established a national policy to promote the environment and also established the President's Council on Environmental Quality. The Act applies to all discretionary actions of federal agencies, including permit approvals of private facilities, which are relevant from a climate policy perspective. The Act does not contain a citizen suit provision. Challenges may only be brought under the Administrative Procedure Act. Nevertheless, to date more than 40 cases have been filed. They seem to be successful mainly in invalidating environmental reviews that failed to consider climate change.

Cases in the United States in which legal standing was an issue are plentiful. *Thrun et al.*,[40] for example, in 2011 challenged the legality of New York's participation in the RGGI system on several grounds, but their causes of action were not addressed substantively because of the absence of legal standing. The Court held that the plaintiffs had failed to establish that their injury was distinct from that of the general public. Utility companies falling within the scope of the RGGI would, however, constitute potentially interested parties. A claimant that was operating a gas-fired combined cycle co-generation facility argued that he suffered hardship because he could not pass on price increases to consumers (*Indeck Corinth*)[41] and did have legal standing in statutory challenges against the RGGI.

In EU law, privileged applicants (such as Member States) and institutions (such as the European Parliament) have extensive rights of legal standing. By contrast, natural and legal persons have limited opportunities to enforce their rights. Direct individual claims in the EU are rejected mainly on the ground of lack of 'direct and individual concern'. Individual concern was evaluated by the Court on the basis of the *Plaumann* test,[42] the criteria of which are considered to be very difficult to meet for natural and legal persons. Cases lodged in the context of environmental law and legal challenges against Commission decisions in the context of the EU ETS were no exception.

Several cases have been brought by private parties challenging various aspects of National Allocation Plans (NAPs) adopted in the context of the EU ETS. The Court has held that the Commission Decisions on NAPs did not affect the legal situation of third parties since it is only when the Member State adopts the NAP that operators are granted rights.[43]

---

[40] *Thrun et al. v Cuoma et al.*, Index No. 4358-11; RJI No. 01-11-104776.

[41] *Indeck Corinth, L.P. v David A. Paterson et al.*, Consent Decree, Index No. 5280-09.

[42] Case 25/62 *Plaumann v Commission* [1963] ECR 95: 'Applicants must show that the decision affects them by reason of certain attributes which are peculiar to them or by reason of circumstances in which they are differentiated from all other persons and, by virtue of these factors, distinguishes them individually just as in the case of the person addressed.'

[43] See, for example, C-6/08 P *US Steel Košice v Commission*, Order of the Court (Sixth Chamber) of 19 June 2008, available at: http://curia.europa.eu/juris/

Clearly the desire to increase individual legal protection under EU Law has led to the strengthening of the legal remedies to challenge EU acts under an annulment procedure (Article 263 TFEU). After the entry into force of the Lisbon Treaty, legislative acts still require plaintiffs to meet the prohibitively high threshold of the *Plaumann* test.

EU acts that are regulatory acts, that do not entail implementing measures, that are of general application, and qualify as non-legislative acts[44] require only 'direct concern'. While regulations and directives *can* be non-legislative acts, decisions will normally qualify as non-legislative acts since they are in most cases adopted on the basis of a non-legislative procedure.

Despite the significant advances regarding legal standing introduced by the Lisbon Treaty, the cases dealing with the legal standing of private parties in proceedings regarding NAPs would not have been decided otherwise today. Article 263(4) TFEU applies only to regulatory acts that do not require implementing measures. A Commission Decision against a NAP would require implementing measures. Consequently, direct and individual concern would still have to be proved. For natural and legal persons, it is very difficult to prove direct concern since their legal situation is not changed as a direct effect of the Commission Decision, but only by the implementation of the NAP itself. Of course, their economic situation may be affected, but they are not deprived of any right they were enjoying earlier. In such cases, direct legal challenges (annulment procedures) against Commission decisions are not possible for private parties, and they would need to rely on action by the Member State.

There remains the possibility of challenging acts of the EU indirectly, via the preliminary ruling procedure under Article 267 TFEU. Article 19(1) of the Treaty on European Union (TEU) now obliges Member States to provide remedies sufficient to ensure effective legal protection in the fields covered by Union law. This provision could oblige Member States to ensure that preliminary rulings questioning the lawfulness of

---

document/document.jsf?text=&docid=67979&pageIndex=0&doclang=en&mode=lst&dir=&occ=first&part=1&cid=824625.

[44] See T-18/10 *Inuit Tapiriit Kanatami and Others v European Parliament and Council of the European Union* (nyr) para. 56 and C-583/11 P, *Inuit Tapiriit Kanatami and Others v European Parliament and Council of the European Union*, Opinion of Advocate General Kokott delivered on 17 January 2013, para. 47, available at: http://curia.europa.eu/juris/document/document.jsf?text=&docid=132541&pageIndex=0&doclang=EN&mode=lst&dir=&occ=first&part=1&cid=822850.

regulatory acts that require implementing measures could be facilitated to ensure the effective legal protection of individuals.

While the Lisbon Treaty has improved the prospects for natural and legal persons to bring claims on the basis of EU Law, legal standing may still be restricted in a large number of cases. During the third trading phase of the EU ETS NAPs do not exist and the Commission has a more prominent role in shaping the nature of the EU ETS. It may be in those situations – where the Commission takes decisions that do not require implementation measures – where the new rules on legal standing may enhance legal protection for natural and legal persons.

In Australia, climate change-related cases have long been litigated even though its emissions trading system is still very young. In *Haughton v Minister for Planning*[45] the legal standing of the plaintiff was denied by the Environmental Planning and Assessment Act 1979 (New South Wales). At issue was the challenge to a Minister's decisions to grant conditional concept plan approvals for power plant projects. The plaintiff relied on the High Court Decision in *Kirk v Industrial Court of New South Wales*,[46] and argued that state legislation could not remove the jurisdiction of the Court to determine and enforce the (jurisdictional) limits of executive power where that power has been exercised in breach of its statutory (jurisdictional) limits. The plaintiff argued, furthermore, that the Minister's decisions amounted to a jurisdictional error. While the plaintiff accepted that the relevant section of the statute applied to non-jurisdictional errors, the provision must be interpreted to accept a challenge that is made against a decision of the Minister on the basis of jurisdictional error of law.

The Court accepted the plaintiff's legal standing and, in doing so, it considered four issues. The first was the existence, under the Environmental Planning and Assessment Act, of a statutory entitlement to start proceedings for any individual. Secondly, the Court noted that the plaintiff had a sufficient right or interest in the matter. Thirdly, it considered that acts and instruments are to be interpreted so as not to exceed the legislative power of the New South Wales Parliament.[47] Lastly, the Court noted the principle that statutes can only bar superior courts from jurisdictions, and the Court accepted the legal standing of the plaintiff. Legal review was therefore granted, even in a situation where, on the basis of the law, legal review was not permitted.

---

[45] *Haughton v Minister for Planning and Macquarie Generation*; *Haughton v Minister for Planning and TRUenergy Pty Ltd* [2011] NSWLEC 217.

[46] *Kirk v Industrial Court of New South Wales* (2010) 239 CLR 531.

[47] Interpretation Act 1987 (NSW), s. 31(2).

Legal standing is not only an issue for natural or legal persons. In the context of climate law and emissions trading systems, stakeholders frequently seek legal standing to litigate on behalf of a group of companies, such as trade associations, or on behalf of a societal interest, such as environmental NGOs.

In *National Chamber of Commerce and National Automobile Dealers Association v Environmental Protection Agency*[48] the Court of Appeals of the District of Columbia Circuit ruled on the legal standing of associations in cases relating to greenhouse gas emissions. The Court held that:

> an organization bringing a claim based on associational standing must show that at least one specifically identified member has suffered an injury-in-fact.... At the very least, the identity of the party suffering an injury in fact must be firmly established.

In relation to establishing 'injury-in-fact', the petitioner must not rely on speculation, but must show that the injury is 'actual or imminent, not conjectural or hypothetical ... and that there is a causal connection between the injury and the conduct complained of'. This rules out a liability regime which is not based on responsibility, and the requirement of 'likely' injury requires some evidence of 'substantial probability'. Consequently, the petitioners' two arguments were not accepted – namely that the motor vehicle standards at issue may result in (i) the automobile manufacturers having to alter the mix of vehicles ('mix-shift') they would otherwise deliver for sale in California, and that (ii) car dealers may lose either customers or profit margins because of the increase in the price of manufactured cars as a result of the standards. The plaintiffs relied on the business plans of motor vehicle companies (Chrysler, for example, reported that it may have to restrict sales as a last resort if the standards were to be introduced), while the respondents relied on government reports. The Court held the latter to be more convincing and substantively accepted the California Air Resources Board's finding that:

> by the time the California standards are fully phased in, any increase in average vehicle costs will be more than offset by fuel-cost savings over the life of the vehicle, resulting in a net decrease in the effective price of automobiles.

48 *National Chamber of Commerce (NCC) and National Automobile Dealers Association (NADA) v Environmental Protection Agency* 42 F.3d 192, 73 ERC 1379, 395 U.S.App.D.C. 193, 29 April 2011.

An association of 'Irritated Residents' that was afforded legal standing to challenge the California Air Resources Board[49] may not have been afforded legal standing in the European Union. As explained above, an important hurdle for legal standing in the EU is the high burden of proof of 'individual concern' embodied in the *Plaumann* test. Since impairment of material interest will be easier to prove than impairment of public interest, interest groups find it particularly difficult to be given legal standing. Interest groups will have standing only if (i) they are granted specific procedural privileges on the basis that recognition by the political process confers an entitlement for protection before the court; (ii) their members are individually concerned; or (iii) in negotiations they are recognized by EU institutions as the central 'interlocutor'.[50] For example, the European steel manufacturing industry trade association (Eurofer) sought annulment of the Commission Decision determining the transitional Union-wide rules for harmonized free allocation of emission allowances,[51] but was denied legal standing on the ground that its members were not individually concerned.

4.2 Legitimate Expectations

Climate change policy is conducted in the context of scientific uncertainty.[52] As scientific understanding of the impacts of anthropogenic greenhouse gas emissions advances, policy changes may be required. It may not come as a surprise that investment plans drawn up against a time horizon exceeding 20 years are to be considered very risky. It is often alleged that 'legitimate expectations' of market operators are infringed; they therefore merit attention. Legitimate expectations are frequently linked with legal certainty. The basis of this concept lies in good faith and requires that, once regulators have induced a business operator to act in a certain way, they should not alter their conduct so that the operator incurs losses.

In *Branco*[53] the Court examined three conditions. First, precise, unconditional and consistent assurances originating from authorized and

---

49 *Association of Irritated Residents v California Air Resources Board* 42 ELR 20127 No. A132165 (Cal. App. 1st Dist., 06/19/2012).

50 See T-585/93 *Greenpeace and Others v Commission* [1995] ECR II-2205; Case T-38/98 *Associazione Nazionale Bieticoltori v Council* [1998] ECR II-4191.

51 T-381/11 *Europäischer Wirtschaftsverband der Eisen- und Stahlindustrie (Eurofer) ASBL v European Commission*, 4 June 2012 (nyr).

52 This passage is based on Chalmers and Tomkins (2007) p. 412.

53 T-347/03 *Branco v Commission* [2005] ECR II-255, para. 102.

reliable sources must have been given to the person concerned by Community authorities. It does not require the applicant to be named; it can also relate to a general statement or a particular course of conduct that encourages expectations. Second, those assurances must be such as to give rise to a legitimate expectation on the part of the person to whom they are addressed. The behaviour of an ordinary and prudent trader will be the objective standard.[54] Third, the assurances given must comply with the applicable rules.

In *Mulder*, under the EU Common Agricultural Policy a farmer was incentivized to cease the production of milk for five years.[55] When he wanted to restart milk production, he was refused a quota on the ground that he had not produced milk in the preceding year. The quota rule was introduced during the five-year period during which he had ceased production. The Court held that when a producer is encouraged by a Community measure to limit his production in the general interest for a short period, he may legitimately expect not to be subject to restrictions that affect him because he took advantage of the Community provisions. This was deemed to be inconsistent with the existing policy framework.

A reversal of policy, however, does not violate legitimate expectations. Rather, there is the presumption of a freedom to legislate because prudent traders should take into account the possibility of legal changes.[56] In exceptional circumstances the Court has been willing to accept legitimate expectations. In *CNTA* the stopping of a subsidy without providing for any transitional period gave rise to immediate and unexpected financial effects and was held to violate legal expectations.[57] After all, even prudent traders would not have been able to avoid the unforeseeable costs.

This line of reasoning of the Court is likely to be the same in the context of emissions trading systems. Consequently, covered entities under the EU ETS will not be able to rely upon legitimate expectations.

4.3 Precautionary Principle

Without being able to go into detail regarding the precautionary principle, it should be mentioned that the principle is increasingly gaining

---

[54] Case 265/85 *Van den Bergh en Jurgens v Commission* [1987] ECR 1155.

[55] Case 120/86 *Mulder v Minister van Landbouw en Visserij* [1988] ECR 2321.

[56] Case 52/81 *Faust v Commission* [1982] ECR 3745.

[57] Case 74/74 *Comptoir National Technique Agricole (CNTA) SA v Commission* [1975] ECR 533.

relevance in case law. The principle was applied in *Greenpeace Australia v Redbank*.[58] Under the Environmental Planning and Assessment Act, Greenpeace Australia appealed against a development consent that had been granted for building a power station in the Hunter Valley. The plaintiffs argued that the power station would unacceptably exacerbate the greenhouse effect and asked the Court to apply the precautionary principle of the National Environmental Protection Agency and to refuse the development consent.

The Court held that the precautionary principle requires cautious consideration of the factors that affected the granting of development consent; however, it does not require the greenhouse gas effect to be given precedence over other considerations. Recognizing other positive environmental and societal effects, the Court did not entertain the plaintiffs' request.

5. CONCLUDING REMARKS

In this chapter a large number of cases relating to climate change issues have been presented. They stem from different jurisdictions, different emissions trading systems and very different areas of law. The chapter has covered a large variety of legal issues, some specific legal areas that are not to be under-evaluated, as well as some critical issues in enacting an emissions trading system. It is therefore not surprising that no detailed conclusion can be drawn. On an abstract level, however, this chapter carries an important message for ETS designers.

It shows that even if emissions trading systems are an economic instrument, the types of legal issue that can arise on their bases are very diverse. Moreover, the angle from which legal issues arise is hard to predict. There is therefore a strong role for the active involvement of lawyers in the design process of an emissions trading system. This role is so far too easily underestimated. Lawyers can help by pointing out potential legal challenges and pitfalls that can arise in the design and operation of an emissions trading system; they have, therefore, an important role to play in the ETS design process.

---

[58] *Greenpeace Australia Ltd v Redbank Power Co Pty Ltd* 86 LGERA 143; 1994 WL 1657428.

9. Linking emissions trading schemes

1. INTRODUCTION

Over the last few decades global concerns about climate change have been intensifying, but only relatively recently have substantial efforts been made to address them. On the international level, the Kyoto Protocol of 1997 sought to forge a global climate alliance. While not succeeding in reducing the concentration of atmospheric greenhouse gas emissions, it gave rise to national and regional steps to curb GHG emissions in a cost-effective way by employing market-based instruments.

One alternative approach to moving forward, and thus combat global warming, is to 'link' existing and future emissions trading systems without having to wait for countries to agree on a global climate change accord. 'Linking' is generally defined as a mechanism that allows market participants in one trading scheme to use emission rights issued under another trading scheme to meet domestic compliance obligations.[1] Linking emissions trading schemes is expected to allow for a wider range of abatement opportunities and a larger market size, which stimulates liquidity and enables a more efficient allocation of resources.[2]

Despite the attention that 'linking' has received in the literature so far, there are very few studies of the legal issues and certainly not from a law and economics point of view, their focus being more on mapping legal barriers than on finding legal solutions. This chapter looks at the possibilities of expanding the benefits of an emissions trading system by linking it to other emissions trading systems. Since design features may determine the ability to link systems, this chapter addresses the question of 'with whom' to establish a trading system. The analysis first sets out the general background for linking (Section 2). It then continues to show the benefits of linking in order to explain why linking is attractive (Section 3). It then assesses impediments that are encountered in the process of linking (Section 4). The practical hurdles that have been

[1] Haites (2003).
[2] Baron and Philibert (2005).

encountered in the linking experiences so far are reviewed in Section 5, and Section 6 concludes.

2. BACKGROUND

The last decade saw a surge in emissions trading schemes as a market-based instrument of pollution control. In the European Union and in Norway Emissions Trading Schemes have been operating since 2005, while Switzerland launched its scheme in 2008. On the other side of the Atlantic, sub-national emissions trading schemes are sprouting in the US and Canada. The Regional Greenhouse Gas Initiative (RGGI),[3] the Midwestern Greenhouse Gas Accord (MGGA)[4] and the Western Climate Initiative (WCI)[5] are the major existing and proposed initiatives for emissions trading schemes in that region. Not only in America but also in the Asia-Pacific region have emissions trading systems emerged. The New Zealand scheme has been operating since 2008. Australia recently passed a bill to introduce an emissions trading scheme starting with a fixed price in July 2012 with transition to flexible market-set prices in 2015.[6] The Republic of South Korea will also introduce a mandatory

---

[3] The RGGI is a sub-national mandatory cap-and-trade programme established in 2009 by ten north-eastern US states (Connecticut, Delaware, Maine, Maryland, Massachusetts, New Hampshire, New Jersey, New York, Rhode Island and Vermont). It regulates CO_2 emissions from power generation in the ten partner states.

[4] The Midwestern Greenhouse Gas Reduction Accord is a regional agreement between six US states (Minnesota, Wisconsin, Illinois, Iowa, Michigan and Kansas) and the Canadian province of Manitoba. It aims, inter alia, to develop cap-and-trade programmes to help combat climate change. The accord has not set any date for commencement.

[5] The Western Climate Initiative (WCI) started as a collaboration between seven US states (Arizona, California, Montana, New Mexico, Oregon, Utah and Washington) and four Canadian provinces (British Columbia, Manitoba, Ontario and Quebec), which have been working together since 2007 to identify, evaluate, and implement policies to address climate change. By December 2011 all Member States from the US except California had withdrawn. California and Quebec have started the first compliance period in their respective cap-and-trade systems. The California Air Resources Board recently set 1 January 2014 as the date on which a link between California and the Quebec cap-and-trade system should be established.

[6] Point Carbon (2011a); Jones et al. (2011).

cap-and-trade system in 2015. China is experimenting with several pilot emissions trading systems and is likely to introduce a carbon tax scheme soon.

Despite the proliferation of national efforts to prevent climate change, global warming continues to be a threat. The recent failure of the global climate change negotiations at the 15th Conference of the Parties in Copenhagen is a keen reminder that the international community is not yet ready to meet this challenge collectively. The Copenhagen negotiations had the ambitious goal of producing a fully fledged, legally binding agreement on climate change. This agreement would lay down a comprehensive framework in order to (i) achieve substantial emission reductions up to 2050, (ii) adapt to the inevitable impacts of climate change, and (c) transfer massive amounts of financial and technological resources from developed to developing countries, allowing the latter to mitigate their emissions and adapt to climate change. With the unproductive ending of these negotiations, a truly global solution has lost much of its appeal for negotiators.

In 2010, at the 16th Conference of the Parties in Cancun, agreement on an international solution to global warming was not forthcoming but the mandate of the Ad Hoc Working Group on Further Commitments for Annex I Parties under the Kyoto Protocol (AWG-KP) was extended. Among the most notable outcomes of the Cancun round was thus the recognition that deep cuts in global emissions were required to limit the increase in the global average temperature to 2ºC and to establish a Green Climate Fund (encompassing US$100 billion per annum by 2020) and a Climate Technology Centre and Network.

At the 17th Conference of the Parties in Durban, governments (including those of 38 industrialized countries) agreed to extend the Kyoto Protocol with a second commitment period lasting from 1 January 2013 to 2020. Thereafter a universal legal agreement on climate change is to take effect; this should be agreed upon by 2015. To facilitate this, the Ad Hoc Working Group on the Durban Platform for Enhanced Action (AGW-DP) was established.

The Kyoto Protocol was amended at the 18th Conference of the Parties at Doha to prepare for its extension. The work of the AGW-DP in paving the ground to reach a global climate change accord is ever more important since several Members of the Kyoto Protocol have not assumed greenhouse gas reduction targets for the second Kyoto Protocol compliance period. The newly forged alliance to prevent climate change thus still appears to be fragile. Linking existing emissions trading systems could therefore be an effective way forward to combat climate change.

3. BENEFITS OF LINKING

Most existing and emerging emissions trading schemes are unilaterally linked to the Kyoto Clean Development Mechanism (CDM). Notwithstanding these unilateral links with the CDM, existing emissions trading schemes remain isolated.[7] When two or more schemes are linked, allowance prices will rise in one market and fall in the other market until full or partial allowance price convergence is achieved. Because the least cost abatement options are spread all over the globe, convergence of allowance prices would deplete these options and lead to achieving an environmental target at the lowest cost possible.[8] Moreover, linking – by enlarging the emissions trading market and creating opportunities for intersystem trading – would lead to a more liquid carbon market with a more stable price signal.[9] Uniform allowance prices would also eliminate competitive distortions that might arise from differences in 'pre-linking' allowance prices.

Linking could also play important roles in the political arena. As climate change is a transboundary environmental problem and the world's economies are becoming ever more interdependent through trade and capital flows, international cooperation lies at the heart of combating global warming. Linking plays three roles in the evolution of such an international climate alliance.

First, a network of links between regional, national and sub-national emissions trading schemes, coupled with unilateral emissions reduction commitments, could give rise to a bottom-up international climate alliance.[10] The proliferation of emissions trading schemes could thus help in paving the way towards a political solution for climate change. This could be an important function at this particular point in time when international interests cannot be reconciled and the establishment of a top-down global climate alliance is becoming increasingly difficult. Linking thereby serves as an important 'second-best' solution in the event that top-down climate negotiations are unable to deliver a timely solution for global warming.[11]

Second, it has been suggested that successful cooperation between emissions trading systems could lead partner nations to value this

---

[7] Tuerk et al. (2009).

[8] Jaffe and Stavins (2007); Tuerk et al. (2009); Flachsland, Marschinski and Edenhofer (2009a); Grull and Taschini (2010).

[9] Jaffe and Stavins (2007); Tuerk et al. (2009).

[10] Aldy and Stavins (2011).

[11] Flachsland, Marschinski and Edenhofer (2009b).

experience and generate more mutual trust. Thus, from a patchwork quilt of different but interlinked systems, a new impetus for global climate change solutions could emerge by emphasizing successful cooperation. In this sense, linking not only constitutes a second-best solution if global agreement cannot be reached, it also serves as a stepping stone in the evolution of a top-down climate policy architecture.[12] It may also be possible that the successful proliferation of one or two types of emissions trading system create a path dependency that increases the costs of not being part of those schemes. In this sense, the extension of schemes once started may prove to be a self-fulfilling prophecy. The adaptations undertaken by Switzerland and Australia (and, to a lesser degree, New Zealand) pursuant to the linking negotiations with the EU ETS would seem to corroborate this point.

Finally, the linking of emissions trading systems allows for cost savings and other economic advantages to be realized. This may incentivize decision makers to provide for links with other emissions trading systems in any future broader climate policy architecture.[13]

4. BARRIERS TO LINKING

Despite the economic and political benefits presented above, linking faces a number of obstacles: the loss of political autonomy will be considered first, followed by lobbying and compatibility issues between climate change policies in different jurisdictions.

Loss of autonomy

Linking results in a form of 'shared sovereignty' in the sense that governments lose some regulatory control over their respective emissions trading schemes.[14] A measure adopted by one of the linking partners transmits itself into, and has repercussions across, the entire interconnected system.[15] Moreover, a country's rate of transition to a low carbon economy is determined to a significant degree by the allowance price prevailing within the overall system of linked emissions trading schemes. This threat of losing regulatory control over some aspects of domestic climate change mitigation policy may limit the political appetite of some governments for linking.

---

12 Aldy and Stavins (2011).
13 Ibid.
14 Granaut (2011).
15 Jaffe and Stavins (2007).

Creating links with other emissions trading systems may be problematic if one is already linked to another scheme. Linking with a more-or-less stringent partner may affect the interests of the linked scheme and may therefore restrict the freedom to establish additional links. If, during the process of the linking negotiations, reform of the scheme becomes necessary, this may have to be done with the consent of the existing partner jurisdictions in order not to endanger the existing cooperation.

Lobbying and rent-seeking

Establishing a unilateral or bilateral link between emissions trading systems will have an effect on covered operators in the jurisdictions concerned. There are two impacts that are important from a policy-making and negotiation perspective. First, linking leads to the convergence of allowance prices across the linked emissions trading systems. This means that in some jurisdictions prices will increase compared to the pre-linking level, while prices in other jurisdictions will decrease. This inevitably creates winners and losers in each of the linking-partner jurisdictions. Where allowance prices increase, operators that are net sellers of allowances will gain. By contrast, net buyers of allowances and consumers of emissions-intensive products or services will lose. On the other hand, the converse holds true in emissions trading systems where linking results in a decrease in the allowance price. By creating winners and losers in the emissions trading systems, linking leads to an intersystem, as well as intrasystem, redistribution of wealth between the various actors. It must be stressed that the general finding of trade theory – namely that trade benefits both economies and that the losses to the losers in each economy are (theoretically) more than compensated for by the winners' gains – hold true. Yet, from a policy makers' perspective the losers in linking are likely to engage in lobbying to prevent linking, while the winners will also engage in lobbying (rent seeking). Such activities can be welfare-reducing and even lead to the creation of (political) barriers to linking.

Compatibility issues

Barriers to linking are not only created through loss of autonomy or lobbying; they can also arise from differences in the design of existing and proposed emissions trading schemes, which vary significantly in size, design characteristics and geographical scope.[16] For instance, provisions relating to the allocation of allowances, offsets, safety valves, and

---

16 Tuerk et al. (2009).

banking and borrowing vary across the range of schemes. Since design characteristics are the results of economic, legal and political factors,[17] resolving these barriers requires the striking of a complex balance between competing policy considerations that underpin the design of the trading schemes concerned. Elements to be considered include: (1) coverage of the schemes; (2) the stringency of the policy target; (3) the nature of the policy instrument; (4) allocation mechanisms; (5) new entrants; (6) *ex post* adjustments; (7) leakage safeguards; (8) offset provisions; (9) banking and borrowing; (10) environmental effectiveness; and (11) standards of privacy and data protection. Each element is discussed below. It bears mentioning that most of these differences obstruct the reaching of a negotiated solution, but they do not constitute an absolute bar to linking because the economic problems to be faced by operators in a linking scenario would also persist in the absence of linking.

4.1 Coverage of the Schemes

Emissions trading systems differ in respect of their coverage. They can, for example, include different industrial and agricultural sectors, or installations in those sectors based on different benchmarks or that address different greenhouse gases. Variations in coverage will lead to different degrees of market liquidity, different operating costs of the emissions trading system and different cost burdens placed on the covered operators. This, in turn, will affect the competitive position of the companies. It will also make the future price movements of emission allowances in the respective trading schemes more difficult to predict.

As was described in Chapter 4, the EU ETS includes stationary operators from five broad sectors: energy (including electric power and oil refineries), the production and processing of ferrous metals (iron and steel), minerals (cement, glass and ceramics), pulp and paper. It also includes the aviation sector and might eventually be extended to maritime shipping. By contrast, the US RGGI system only covers CO_2 emissions from around 200 fossil fuel-fired power plants; other sectors are therefore not covered. In the RGGI system it is only emissions from fossil fuel-fired power plants that are larger than 25 MW that are included; thresholds for installations in the EU ETS are largely in the area of 20 MW. Small installations can be excluded from the EU ETS scheme if

[17] Behr, Witte, Hoxtell and Manzer (2009).

they emit less than 25,000 tonnes of carbon dioxide equivalent and where they carry out combustion activities with a rated thermal input below 35 MW.[18]

Similarly there are differences in the types of greenhouse gas covered by the scheme. In Australia it is only carbon dioxide, methane, nitrous oxide and perfluorocarbons that are covered, while in other schemes (the EU ETS, New Zealand and South Korea, for example) all six major greenhouse gases are covered by law.

4.2 Stringency of the Political Target

The political acceptance of climate change responsibility varies strongly among jurisdictions. The EU, for example, aims to reduce greenhouse gas emissions by 20 per cent below 1990 levels in 2020, or even by 30 per cent if other major economies concede to do their fair share of preventing climate change. By 2050, the EU seeks to have reduced its emissions by 80 to 95 per cent below 1990 levels. Politicians in Australia aim to reduce emissions by 5 per cent by 2020 and by 80 per cent by 2050 below 2000 levels. Different reference dates for setting benchmarks will lead to varying degrees of stringency among jurisdictions.

China has recognized climate change as an important challenge. Unlike the EU, which sets its targets on the basis of absolute emission values, China uses intensity targets. In its 12th Five-Year Plan, China has set a 17 per cent reduction target of carbon intensity per unit of GDP below 2005 levels by 2015.[19] This will allow China to make important progress towards honouring its commitment to reduce the carbon intensity per unit of GDP by 40 to 45 per cent compared with 2005 levels by 2020.

The setting of different targets can complicate the assessment of the relative stringency of the various existing or emerging emissions trading systems. The stringency of the schemes will directly affect the price of emissions allowances and thus the competitiveness of the covered industries. Such differences in the chosen targets will make it more difficult for negotiators to assess the relative impacts of the linking process on the various stakeholders.

---

18 Directive 2009/29/EC, Art. 27.

19 Sandbag (2012) p. 12.

4.3 Nature of the Policy Instrument

Jurisdictions can choose among a variety of instruments to address climate change. They can, for example, use an emissions trading scheme based on cap and trade, a tax scheme, a price collar scheme (that is, a cap-and-trade scheme that operates within the bands of a price floor and a price ceiling), or they can select a performance standard rate (PSR) system that operates as a credit trading scheme in the absence of a fixed emissions cap. When only a tax scheme is employed, linking to such a system will not be possible as there will be no emission allowances to be used. The linking of a cap-and-trade scheme with a price collar scheme and a PSR scheme is examined below.

A study in 2009 promoted the idea of a price ceiling based on McKibbin and Wilcoxen,[20] and paired it with a price floor as being an efficient way of linking emissions trading systems.[21] The underlying idea is a system of price collars in which a trading system could fluctuate. The narrower such price collars are, the more comparable they will be to taxation systems that might evolve in other jurisdictions. Such a system may make the effort required from operators in both jurisdictions more comparable since the price variation is more limited than it would be under a cap-and-trade scheme.

A price collar system of this type was planned to be introduced in Australia in 2015 (to run from 1 July 2015 to 30 June 2018). Until this date the Australian Carbon Pricing Mechanism (CPM) will operate on the basis of a fixed-charge scheme that is comparable to a tax scheme. As part of the still ongoing linking negotiations with the EU, Australia has had to abolish its price floor and bring its scheme closer to the European scheme.[22]

The price floor in the Australian CPM was to be set at AU$15 per tonne of CO_2 equivalent for 2015–16 and would increase by 4 per cent annually in real terms to adjust for inflation.[23] The price floor would have been operated by an auction reserve price for Australian Carbon

---

[20] McKibbin and Wilcoxen (2002).

[21] PriceWaterhouseCoopers (2009).

[22] The Clean Energy Amendment (International Emissions Trading and Other Measures) Bill 2012, containing the abolition of the price floor, was approved by both houses of Parliament on 26 November 2012.

[23] See Commonwealth of Australia (2011) p. 104.

Units,[24] as well as a surrender charge for eligible international emissions units.[25] The surrender charge would have resembled a levy in the sense that it sought to inflate the price of international emissions units towards the Australian price floor.

Although the price floor has been abolished, the price ceiling still remains. The price ceiling will be set in 2014 by regulation at AU$20 above the expected international price for 2015–16 and will rise by 5 per cent per year in real terms.[26] If the price ceiling is triggered, the regulator is to issue fixed-charge carbon units that are similar to those for the fixed-charge period (from July 2012 to June 2015).[27]

Linking a price collar system with a cap-and-trade system may have practical implications. To elucidate these implications we assume that Australia still has a price collar system, as was foreseen, and will establish a unilateral link with the EU ETS. This means that EUAs may be used for compliance purposes in Australia. If, for one reason or another, the price of allowances in Australia were to fall below the price floor, the country would be obliged to support the price of allowances in Australia. This would be done by means of an auction reserve price or via a levy on offsets. In a unilateral linking situation, Australian operators are able also to use EUAs for compliance, but the EUAs would not be treated differently from other offsets. This implies that there would be only limited demand for EUAs when the price floor would have been triggered. Given the current oversupply in the EU ETS, it is therefore not surprising that the European Commission requested the abolition of the price floor.[28]

It bears mentioning, however, that the Australian price ceiling is still in place. In the framework of a bilateral link between both systems, a triggering of this price ceiling would lead to a supply response from the government that would also affect the supply available to operators

[24] An Australian Carbon Unit is a domestic carbon currency issued by the CPM's administrator and represents a tonne of CO_2 equivalent greenhouse gas emissions: see Department of Climate Change and Energy Efficiency (2011) p. 91.

[25] Ibid., p. 91.

[26] See Commonwealth of Australia (2011) p. 104. The price of European Union Allowances (EUAs) will be taken as a reference to set the price ceiling: see http://www.cleanenergyfuture.gov.au/wp-content/uploads/2012/08/CEF-FS43-Linking-liable-entities.pdf (last accessed 11 December 2012).

[27] Department of Climate Change and Energy Efficiency (2011) p. 91.

[28] Another explanation can, of course, be found in the European Commission's aversion to any 'price setting' or 'price targeting' in the context of the EU ETS.

covered by the EU ETS. It would directly ease the supply conditions for EU ETS operators and thereby limit EUA price increases in the event of a bilateral link. In this respect, the European Commission does not yet appear to be concerned.

In addition to a price collar system, policy makers could also choose, for example, a credit-and-trade trading system such as a performance standard rate (PSR) system, under which entities are free to emit but must compare their actual emissions level with a government standard. A company earns emission credits if it over-complies with the standard, or will need to purchase additional credits if it falls short of it. The emitting entity has to account for every emitted tonne in excess of the standard and pay a penalty if it cannot attain the credits. Unlike the EU ETS system, with the PSR system emission allowances are not created by the government and distributed to emitting entities, but are created automatically on the basis of the law. A legal enactment obliges emitting entities to meet particular emissions standards, to have them verified by third parties and to report them to the government. While the government takes an active role in monitoring, it does not participate as an actor; it takes a *laissez faire* approach, limiting itself to creating the rules.

Because a PSR system does not have a binding emissions cap, it cannot be ensured that price developments within such an emissions trading system will be comparable with those under a cap-and-trade system. Cost differentials may reduce the competitiveness of those entities that have to bear the higher price, although it must be stressed that this comparative disadvantage would persist even in the absence of a link between the trading systems. Linking may be particularly useful if negotiations were to lead to the introduction of an emissions target in the PSR system that would trigger the continuous adjustment of the trading system so as to automatically adjust the emissions standard in order to create additional scarcity in the credit market. This would help to limit the price differentials between a cap-and-trade system and a PSR system.

4.4 Allocation Mechanisms

Different allocation mechanisms can give rise to different distributional impacts in the linked emissions trading schemes. Clearly, if operators in one trading scheme receive allocation on the basis of historic emissions (grandfathering) while comparable operators in the other scheme must pay for their allowances at auction, the latter group will be at a comparative disadvantage – not, of course, because the market price of the allowances will be different but because they will need to reserve cash for the purchase of the allowances. Similarly, competition between operators could be distorted if auction proceeds were earmarked for investment in adaptation, transformation of the energy sector or channelled back to covered industries, rather than becoming part of the general budget. Such distortions would, of course, also arise in the absence of the link between the trading systems, but they may constitute a contentious political issue within cooperation on emissions trading. Allocation mechanisms may need to be harmonized in order to make linking between two emissions trading systems politically viable.

It is also conceivable that particular allocation mechanisms favour a particular group of stakeholders in an emissions trading system, which may benefit from linking precisely because they are allocated allowances on a favourable basis or because they would benefit from lower compliance costs as a result of linking. If the winners in a linking agreement are successful in their rent-seeking activities, it may not be necessary for the majority of the covered actors to be in favour of linking. In such a scenario, the differences in the allocation mechanism may not only prove to constitute an impediment for linking but also function as a catalyst for linking in the sense that it favours a group that is well-organized and capable of influencing policy-making.

An example of the situation where different allocation formats prevent the equalization of abatement costs arises when free allocations for future trading periods are based on the emissions or output of an entity in the current trading period (updating of allocation base). In this situation covered entities must take into account the reduction in future allocations when deciding to abate or reduce output. Such production subsidies[29] should be considered when examining the effects of linking scenarios. Evidently such subsidies that influence the increase in the marginal abatement costs also distort the abatement behaviour of the emissions trading systems: the system with the updating policy will abate less.

4.5 New Entrants

The treatment of new entrants has been subject to much discussion in the framework of emissions trading systems, as they must be designed to allow new operators to enter the market and to compete with the existing incumbents. If new entrants are not able to acquire emission allowances on the terms on which the incumbents have received them, they are at a comparative disadvantage. If new entrants do not have unqualified access to financial markets they may not be able to enter the market at all. The

[29] Hepburn et al. (2006).

treatment of new entrants can have two types of repercussion that would be likely to affect linking negotiations. Firstly, if one jurisdiction treats new entrants more favourably than does the other jurisdiction, all else being equal, new entrants are more likely to enter the market where they are treated more favourably. Secondly, if undeleted new entrants reserves are made available to the market at the end of a trading period, supply will be larger and allowance prices will be lower. This could lower the market price of allowances in both emissions trading systems. This could involve gains and losses to stakeholders, potentially lead to a weakening of the environmental effectiveness of the trading system and generate repercussions on other (energy) policies that require a sufficiently high CO_2 price to incentivize investment.

4.6 *Ex Post* Adjustments

Ex post adjustment is a broad term with different meanings in different contexts. It can take place at: (1) operator level and relate to adjustments that only marginally impact on the overall amount of allocation and hence the price level, or it can (2) work as a safety valve for the overall emissions trading scheme when market prices are deemed to be too low to stimulate investment or if the environmental integrity of the scheme is endangered.[30] Another way (3) in which *ex post* adjustments are used is to correct the number of allowances generated by offset projects. Since the latter occurs outside an emissions trading system as such and merely determines overall supply of offsets, it is not considered here. The first two *ex post* adjustments are considered in relation to their impact on linking emissions trading systems.

In the parlance of the EU ETS, for example, *ex post* adjustment is often used to describe the recouping of allowances that were allocated for free from those operators that were not using them as a result of a downsizing of capacity of the installation or its complete closure. This was a much discussed topic during the first trading period. Now, in the third trading phase, a larger proportion of allowances is auctioned and hence the problem has diminished. A benchmarking system that takes into account historical production levels could still be subject to *ex post* adjustment in the context of plant closure or downsizing.

---

[30] *Ex post* adjustments can, of course, also be designed to increase the number of allowances available on the market, for example, in cases when the cost burden on the covered sectors is deemed to be too high. Trigger prices or price ceilings, discussed in Chapter 4, are an example of such *ex post* adjustments.

From a political perspective it should be mentioned, however, that the European Commission took a very strong stance to prevent any form of *ex post* adjustment. From a linking perspective it may be expected, therefore, that the Commission is unlikely to become excited about establishing a link with a scheme that contains *ex post* adjustments. In general, however, closure rules are unlikely to have a significant effect on the overall quantity of allowances and hence should not constitute a major obstacle to linking.

The necessity for large-scale *ex post* adjustment can also arise from systemic problems with the emissions trading scheme. In cases of allowance oversupply, for example, market prices may be unduly low to stimulate green investment in the economy. These situations can result from over-allocation or from an economic downturn. The EU ETS is currently suffering from a significant oversupply situation resulting (in part) from the economic depression in the EU. The European Commission seeks to address this situation by, first, backloading 900 million tonnes of CO_2 emissions to 2019 and 2020 and then requesting the legislator to enact new rules in order to cancel these allowances. As mentioned above (Chapter 4), the European Parliament has voted against this option, but a new vote is expected to take place in summer 2013.

Systemic *ex post* adjustments could also be motivated by environmental integrity concerns of the emissions trading scheme. If a PSR system were to be subject to a particular climate policy target, for example, the legislator would have to fine-tune the performance standards in order to achieve its desired target. *Ex post* adjustment here is particularly sensitive and decisions to adapt the performance standards would probably not be designed to have retroactive effect. They might still be qualified as *ex post* in the sense that they may affect the current developments of the emissions trading system during a particular trading period and hence affect a linked trading system unexpectedly.

Large-scale *ex post* adjustments in one emissions trading scheme will also have direct effects on the linked scheme. The reduction in allowances available in one scheme would impact upon demand and supply patterns and hence the price levels in both trading systems. Different rules on *ex post* adjustment can thus lead to a compatibility problem for emissions trading schemes, to shifts in the competitiveness of the respective industries and to political resentment.

4.7 Leakage Safeguards

Leakage safeguards are measures provided by the legislator to prevent de-industrialization (that is, the migration to other states of production, employment and labour). In the literature de-industrialization at modest allowance price levels does not seem to be a major concern. It may, however, still be an issue for linking emissions trading schemes. Competitive concerns could arise for competing operators based in different jurisdictions and which are subject to policy measures that differ in terms of the cost burden placed on covered entities. This may lead to welfare-reducing lobbying activities in both jurisdictions in which policy makers of both countries agree to give more and more support to their industry. This support is, however, positional to the extent that the support schemes are directed towards each other. This means that the effect of the domestic measures taken is partially offset by the measures taken within the foreign system.

In the Australian CPM emission-intensive trade-exposed (EITE) sectors will receive between 60 and 94.5 per cent of the allowances they require for free.[31] Similar measures are available in the New Zealand scheme. Here agriculture and high EITE industries can receive up to 90 per cent free allocation per unit of production; moderate EITE sectors will receive around 60 per cent. In the EU ETS sectors and sub-sectors exposed to a significant degree of carbon leakage are eligible to receive up to 100 per cent of allowances for free.[32, 33] Allocations are, in general, based on *ex ante* benchmarks based upon the average of the top 10 per cent most efficient installations in a sector or sub-sector during the period 2007–08.[34] In addition, the European Commission has adopted guidelines under which Member States can compensate some electro-intensive users

---

31 The actual number of emission allowances an EITE company receives in Australia depends on (i) trade exposure, (ii) the production level of the current year, and (iii) the historic industry average emissions base line for the relevant activity.

32 Directive 2009/29/EC, Art. 10a(12).

33 The number of allowances allocated to a particular installation is determined as follows: benchmark × historic activity level × carbon leakage exposure factor × cross-sectoral correction factor or linear factor. Benchmarks are product-specific based on the top 10% most efficient installations in 2007–08. This benchmark remains valid until the end of the current trading phase. The historic activity level is calculated by either taking the average of the two median production values between 2005 and 2008 (thus excluding the highest and lowest value during this period) or by taking the average of the production levels in 2009–10; the higher of the two will be taken. The carbon leakage exposure factor is set at 1 for EITE industries and at 0.8 for non-EITE industries (the electricity sector is generally excluded from free allocation). The latter will decrease over time to 0.3 by 2020; thereafter it is intended to be reduced to zero in 2027.

34 Directive 2009/29/EC, Art. 10a(2).

for part of the higher electricity costs that are expected to result from the changes to the EU ETS in the third trading phase. The guidelines allow the granting of subsidies of up to 85 per cent of the cost increase faced by the most efficient enterprises in each sector from 2013 to 2015, a cap that will gradually fall to 75 per cent in 2019–20. The effects of these guidelines will be very hard to predict for a potential linking partner since the aid may be granted is not dependent on Union measures but subject to the discretion of every Member State. Leakage safeguards appear to be a uniform concern for policy makers across the globe. Since different jurisdictions have found different approaches to deal with de-industrialization, carbon leakage is an important point to assess when linking different emissions trading systems.

4.8 Offset Provisions

In principle, the different definitions of allowances could give rise to compatibility problems between linking systems. So far it appears that the allowances in most systems are defined in terms of metric tonnes of CO_2 equivalent. Since they mostly share a common denominator for transferring allowances from one system to the other, these transfers are generally possible. Incompatibility between allowances is not based on the common denominator but is related to (a) the source from which they derive, (b) quality concerns, or (c) overall quantity restrictions. The emissions trading systems in Australia and the EU will be taken as an example to demonstrate these points and to show their effects.

Restrictions based on the source relate to the difference in coverage of offset rules. The European rules, for example, do not allow for any forestry-related credits. By contrast, the Australian system accepts credits from its domestic offset scheme known as the Carbon Farming Initiative (CFI). In a potential linking scenario between Australia and the EU, operators in Australia could sell their domestic allowances and use (cheaper) CFI credits for compliance. This would result in lower compliance costs and distortions in competition. Even though such distortions would be present in the absence of linking, it may affect linking negotiations. Australian operators will thus be able to meet their respective obligations with offsets that cannot be used by operators covered by the EU ETS; they would thus be able to meet their compliance obligations at a lower cost and would benefit from selling Australian Carbon Units (ACUs) to the linked scheme. They would thus enjoy a competitive advantage.

Limits on the importation of offsets into an emissions trading system ensure that at least part of the greenhouse gas reduction is realized domestically. If the utilization limits vary significantly, linking could lead to a so-called 'back door problem' whereby a tighter utilization limit of one scheme will be compromised by the less stringent rules of the other scheme.

4.9 Banking and Borrowing

Banking and borrowing, as discussed earlier in Chapter 3, are widespread features of existing and emerging emissions trading systems. Banking allows covered entities to accumulate emission allowances over time and to use them for compliance, when needed. Borrowing allows the use of emission allowances before they are allocated. Both features facilitate the intertemporal smoothing of allowance supply and demand patterns.

Most emissions trading systems allow for banking within one emissions trading phase. The motivation for limiting the banking period to a particular allocation period is because at the beginning of an allocation period the emissions trading system can be recalibrated. In the event of over-allocation for a particular trading phase, the legislator can allocate fewer allowances and thereby create more scarcity. What would be the implications of linking with a system that – like the New Zealand system – allows for unlimited banking? A link between New Zealand and an emissions trading system that allows banking only within a given time period leads to the undermining of the banking requirement. Towards the end of the trading period, prices between New Zealand and other allowances would start to diverge since the New Zealand allowances would not lose their value at the end of the period. Any allowance holding converted into New Zealand credits could be used for compliance after the start of the new trading period. This is, of course, attractive for covered entities in both jurisdictions, but complicates the task of the policy maker in ensuring an adequate level of scarcity in the market.

Borrowing is not a widespread feature in emissions trading systems. Many systems do not allow any borrowing at all, or do not allow interphase borrowing, or only allow borrowing to a limited extent. Australia allows borrowing of up to 5 per cent of an operator's compliance liability. In New Zealand, by contrast, no borrowing is possible at all. Linking between those two jurisdictions would increase the supply in New Zealand via borrowing in Australia. If New Zealand prices were to be temporarily very high, Australian companies could borrow and sell those allowances to New Zealand. The practical effect of borrowing such reduced amounts of allowances permitted by Australia will be limited, particularly since the New Zealand system has lenient rules governing offsets. This may suggest that different borrowing requirements may not always constitute a significant hurdle for linking negotiations.

4.10 Environmental Effectiveness

Environmental effectiveness ranks among the prime concerns when considering linking emissions trading systems. Environmental effectiveness, particularly in a cap-and-trade framework, could be impaired when a link is created with a system that has different design features. These features are offsets, price ceilings and linking with an emissions trading system that is not based on cap and trade. All issues have been discussed above from an economic point of view; here, they are reviewed briefly from an environmental perspective.

If an emissions trading scheme has less stringent rules on offsets and allows types of offset that are not allowed in the other scheme, a link between the two implies that the environmental integrity of the second system is undermined. Operators will use 'low quality' offsets for compliance and sell 'high quality' allowances to the linking partner operators. As a consequence, the use of 'low quality' offsets spills over indirectly to the other emissions trading system.

Environmental integrity is also endangered if a link is created with an emissions trading system that uses a price ceiling. If the price ceiling is triggered, the government issues additional allowances. Since these allowances can also be used for compliance in the linked scheme, the predetermined emissions cap is undermined.

Similarly, the environmental integrity of a cap-and-trade system may be undermined if a link is created with a system that does not operate on the basis of a cap. Economic growth in such a system could increase the overall number of allowances available in the partner jurisdiction. These can also be purchased by operators in the linked emissions trading system and consequently extend the overall quantity of allowances beyond what is available on the basis of the emissions cap.

4.11 Privacy and Data Protection

Differences in privacy and data protection regulation may make it difficult for emissions trading systems to exchange data. Exchanges of data on allowance trades, however, are essential to ensure the interoperability of linked emissions trading systems. Differences in privacy and data protection standards could therefore constitute an impediment to linking, particularly if one of the linked systems establishes further linkages with other trading systems.

5. PRACTICAL EXPERIENCES

So far, there is little experience with the actual linking of emissions trading systems. Relatively little is known about the negotiations since they are, of course, conducted behind closed doors. This section is therefore based upon publicly available information regarding the linking negotiations of the EU ETS with the Swiss ETS and the Australian CPM. It presents the current state of the negotiation processes and highlights the main insights.

The first attempt to link an emissions trading scheme was made by Switzerland and the EU. In light of Switzerland's EEA membership and the small scale of its emission trading system, onlookers at the linking negotiations might have expected a quick conclusion of the negotiations. The initial exploratory talks between Switzerland and the EU on linking their systems date back to 2008, and the negotiations are still ongoing. Following a proposal by the European Commission, the Council of Ministers in December 2010 granted the mandate for commencing negotiations on linking the Swiss and the EU systems. The Swiss system, with 6 million tonnes of CO_2 and 50 covered entities, is significantly smaller than the EU ETS, which covers around 12,000 installations with 2 billion tonnes of emission allowances. Despite the very small scale of the Swiss emissions trading system, three negotiating rounds have taken place so far to prepare the way for a bilateral linking between the two schemes and the creation of a common market for emission allowances. A full revision of the Swiss emissions trading law, which entered into force on 1 January 2013, made the Swiss system largely compatible with the EU ETS. The linking negotiations extend not only to stationary installations but also to the aviation sector and cooperation in technical areas such as the emission allowance registries.

The EU also wishes to link its ETS with Australia. Both seek to conclude an agreement on establishing a full link of their carbon markets; this linking agreement should be adopted by mid-2015 so as to facilitate the start of the fully fledged link by 1 July 2018. During the period 1 July 2015 to 1 July 2018, the Australian CPM will be unilaterally linked with the EU ETS, meaning that only Australian entities are able to use EUAs for their domestic compliance.

This extensive linking agreement between the Australian CPM and the EU ETS should cover a multitude of topics. These include provisions on monitoring, reporting and verification, and also on market oversight. Stipulations on the role of land-based domestic offsets and third party emission units that may be accepted into both schemes will be included.

Issues covered regarding third party allowances will relate, inter alia, to the types and quantities of permissible allowances. Quite importantly, the agreement will also address competitiveness issues and carbon leakage.

In the linking negotiations Australia had to agree to a number of amendments to its emissions trading system: it had to drop its price floor, set the price ceiling for 2015–16 with reference to EUAs and lower the restriction on the use of certified emission units (CERs), emission reduction units (ERUs) and removal units (RMUs) by 12.5 per cent.

6. CONCLUSION

The experiences on linking emissions trading schemes sketched above suggest a number of important insights. They show that linking negotiations with a large emissions trading system, such as the EU ETS, may lead to the proliferation of its key design features. When designing an emissions trading system, designers are therefore well advised to consider possible linking partners early in the design process. This avoids costly adaptations to the system at a later stage.

Key emission trading design issues are by no means the only elements that need to be agreed upon. Clearly such design issues will be addressed at an early stage in a linking negotiation process, but systems must be compatible at a much higher level of detail. The practical rules on implementation (monitoring, reporting and verification) must be, and must remain compatible in order to allow for the development of a level playing field and to prevent distortions in competition and carbon leakage. It also bears mentioning that linking requires adaptation of the registry system in both emissions trading systems to allow for the efficient exchange of allowances.

Perhaps as equally important as establishing the link between emissions trading systems is that the parties must include provisions and procedures on how to deal with future extensions between systems. It is conceivable, for example, that Australia would like to link with the Californian Emissions Trading System. It must be clarified how this would affect the EU ETS, how it would affect the agreement currently being negotiated, and how an emissions trading system would be able to de-link again. Linking, therefore, is a very complex and perhaps greatly underestimated difficult process where 'the devil may be found in the detail'.

10. Concluding remarks

1. INTRODUCTION

Climate change has been on the agenda of national governments and the international community alike. For more than two decades the international community has been deliberating and seeking effective ways to address global warming. Despite some progress made by the members of the Kyoto Protocol (Annex I countries), the concentration of greenhouse gases in the atmosphere continues to increase.

Taking climate change action requires the setting of a clear international target at the level of which global temperature increases should be halted. This is complicated by the inherent (scientific) uncertainty of the topic, further obstructed by collective action problems. Collective action problems relate to 'free riding' – that is, the understanding that if everybody reduces its greenhouse gas emissions, there is always one that can get away without taking action. While this has an intuitive appeal, the actual situation is, of course, more complex.

The various national actors share different historic responsibilities for the escalation of greenhouse gases in the atmosphere and at present contribute in different degrees to climate change. For example, because of its late industrialization, China has little historic fault in climate change, but is currently the largest emitter of greenhouse gases. China is also a vast country and the per capita carbon footprint of one of its residents is much lower than that of a person living in central Europe or even the United States. Similarly, the per capita income is still low and people strive to improve their living standards. Since China will also be affected by global warming, it has assumed responsibility by setting intensity targets. The international community has recognized the principle of 'common but differentiated responsibility', which can help more countries to take action.

Several countries have recognized emissions trading systems as a cost-effective means of reducing greenhouse gas emissions and have started to introduce such systems. When setting up such a trading system, policy makers face similar design challenges. It was our ambition to present insights into ETS design in an easily accessible way and to reflect critically upon the implementation challenges. We undertook to focus more attention on the design challenges rather than present stylized typologies of trading systems and their merits and demerits. By recognizing that policy makers will frequently pursue several – at times conflicting – policy objectives, we hope to help them to raise the right questions when approaching the literature to find the best 'ETS design' for their jurisdiction and the tasks that they want it to perform.

The main issues of this book are therefore how to design an emissions trading scheme, the potential problems in implementing the system, and how they can be addressed.

The following sub-questions were examined:

- What are the main advantages and disadvantages of emissions trading in general?
- What needs to be considered when designing an emissions trading system?
- What are the main implementation problems of emissions trading?
- How should those implementation problems be addressed in an effective, efficient and acceptable way?

Throughout the text those sub-questions were translated into four more simple questions to guide the reader: why, how, what, and with whom? The book thereby sought to assess the advantages and disadvantages of emissions trading (why?) and to examine ETS design variants and policy-making choices (how?). The question 'what?' was addressed by examining concrete implementation problems in ETS design, highlighting important challenges and how they could be addressed in an effective, efficient and acceptable way. The last question 'with whom?' is addressed by examining the issue of linking emissions trading systems.

The remainder of this chapter will therefore present the main insights of this book by asking why? (Section 2), how? (Section 3), what? (Section 4) and with whom? (Section 5).

2. WHY?

Chapter 2 examined how various policy instruments (command and control, liability rules, taxation and emissions trading) compare with each other and particular attention was paid to their strengths and weaknesses. It was shown that emissions trading is attractive because (i) its environmental effectiveness can be safeguarded by ETS design; (ii) greenhouse gas abatement is cost-effective; (iii) it automatically adjusts for inflation;

(iv) political acceptability is high; (v) prices are determined by the market (automatically stabilizing the economy); and (vi) it can incorporate transfers to developing countries.

Since all instruments presented have advantages and disadvantages, it was found that no single instrument is optimal to internalize environmental externalities in the most effective way. In practice, this means that for a particular problem a case-specific solution must be employed. The question that policy makers should be asking is not which instrument should be chosen to address climate change, but rather how the optimal mix of instruments should be assembled for their jurisdiction.

Of particular relevance for ETS designers is the fact that an emissions trading system can deliver cost-effective abatement of greenhouse gas emissions. It incentivizes installations with the lowest abatement costs to abate without the need for policy makers to be aware of the particular abatement costs of individual firms. To take advantage of the appealing properties of emissions trading systems, low administrative costs are important. This is where emissions trading fares less well than simple tax schemes. It is possible, however, to design a system in which the benefits of a wider coverage of abatement opportunities outweigh the additional costs associated with the increasing administrative burden of including more (and smaller) operators. Another approach to limiting administrative costs is to cover different greenhouse gases or (initially) include only fuel input emissions rather than process emissions. ETS designers will need to strike the balance on the basis of a cost–benefit analysis.

In light of an 'optimal instrument mix', it may be attractive to employ a CO_2 tax scheme for smaller emitters and an emissions trading scheme for larger installations. This enables policy makers to introduce greenhouse gas abatement incentives while keeping administrative costs low. This solution is, in fact, applied in Switzerland.

Applying a tax scheme and an ETS scheme to identical installations at the same time does not appear to be desirable. It would increase administrative costs for the affected installations and hence be inefficient if the same results could be achieved by a single instrument.

3. HOW?

The question 'how' to design an emissions trading system is approached by taking a practical turn. The text does not present academic typologies of design variants and does not conclude that a certain design is superior to another – after all, such findings are based on particular assumptions that may or may not be true in the particular context in which the emissions trading system will be implemented. The approach taken in Chapter 3, therefore, was to highlight the importance of the 'context' within which a system is to operate. Recognizing that policy makers typically will have more than one policy objective that they wish to achieve with their ETS design, examples of policy objectives were reviewed.

The chapter then examined how these policy objectives could be reached in an environmentally effective, efficient and politically acceptable way. It became evident that the best ways to achieve these objectives would prescribe different courses of action. Policy makers are thus forced to make trade-offs between them. Even though it is, of course, understandable that policy makers have multiple goals and balance the various policy objectives against each other, this approach has its drawbacks. The policy instruments may become more complex than that which is necessary and thereby become less efficient. According to Nobel Prize Laureate Jan Tinbergen, each and every policy target needs to have at least one policy tool (the Tinbergen Rule).[1] This suggests that overcharging policy instruments with multiple goals is undesirable from an economic perspective.

Chapter 3 also presented the various ETS building elements that designers can use to construct their ETS. Their merits in terms of environmental effectiveness, efficiency and political acceptability were sketched. We hope that this gives ETS designers a productive overview of the choices at their disposition and allows them to study their design choices in greater detail and with the right questions in mind. One of the most important messages of this chapter, therefore, was that policy makers must be very clear about their policy objectives in order to be able to design their system accordingly.

Chapter 4 followed up upon the approach taken in the previous chapter by presenting existing emissions trading systems. ETS designers will appreciate the strong diversity that exists between the various systems, particularly in the area of target setting and the types of covered gases. This makes a direct comparison between the systems difficult, particularly when ETS designers seek to assess the competitive effects of one of those schemes upon their domestic industry. Despite the diversity, there are common approaches. Most systems employ industry support schemes for energy-intensive trade-exposed sectors or, in one way or the other,

[1] See Knudson (2009).

seek to limit the costs of the trading system upon their industry (for example, by increasing the supply of allowances when trigger prices are reached).

4. WHAT?

This question asks what are the concrete implementation problems of which ETS designers should be aware when designing their system. Chapters 5 to 8 highlighted important challenges and how they could be addressed in an effective, efficient and acceptable way.

Chapter 5 examined allocation issues by reminding policy makers that the setting of an environmental target is followed by allocation of emission allowances. Emission allowances must be limited in number – if they are not there will be a situation of 'over-allocation', which defeats the object and purpose of an emissions trading system. It must be avoided at all costs. Besides the over-allocation of the primary allocation of allowances, there is the related problem of oversupply. Oversupply can arise, for example, from a downturn in economic activity, as well as from a strong inflow of offsets into the emission trading system. Another example of a shock that could give rise to oversupply is a reduction in emissions as a result of a change in the energy production mix; this could come, for example, from a strong increase in the use of shale gas.

A further issue that could arise in the context of free allocation is the political problem of 'windfall profits'. Operators that receive allowances (for free or otherwise) have incentives to pass their costs on to their consumers. Their ability to do so depends on their bargaining power. Since energy consumers cannot easily reduce their energy consumption, they have to pay higher prices even when energy producers have received their emission allowances for free. Consumers and politicians in the European Union saw this as unfair and the system of free allocation based on grandfathering was changed to auctioning.

Lobbying is an inherent feature in the design process of any instrument. The design of an emissions trading system is unlikely to be an exception. ETS designers should be aware that stakeholders seek to advance their own interests. Policy makers should be aware of the inefficiencies that can arise from the various design choices that are favoured by the various stakeholders and resist the temptation to give in to those demands that lead to inefficiencies. If ETS designers need to make a trade-off – for example, to muster more political acceptability for the emissions trading system – they should be aware of the societal costs and benefits. Only an objective assessment of those effects can yield an optimal emissions trading design.

Chapter 6 then focused on the secondary market. The effective functioning of the emissions trading market is often taken for granted. This should not be done. An emissions trading market is not so different from other markets. They need adequate rules and market oversight to function efficiently. The chapter reviewed a series of situations in which the secondary market could be subject to manipulation and abuse, and hence not function well. The first element concerned market manipulation. It is often assumed that because the EU ETS market is large, manipulation is not possible. This is erroneous. Financial markets do not require dominance in the industrial economic sense in order to be able to execute market power and to influence prices. The size evidently makes market cornering more difficult – but not impossible. Small production companies, in particular, fear that a market may be manipulated. Large market players, by contrast, argue that manipulation does not occur. ETS designers are encouraged to take this problem into consideration when starting to design their system and not to wait for as long as the European legislator did to address this issue.

Another element that was examined was whether abusive practices could be employed in the context of an ETS auction. Abuse can occur in the absence of the industrial economic concept of dominance and take the form of signalling or demand reduction. Both can be addressed by the adequate choice of auction mechanism designs. If ETS designers fail to take this up (for example, in order to have a simpler auction design format that attracts more bidders and thereby strengthens the competitiveness of the auction market), then efficient and effective market oversight structures must be established. The EU legislator failed to do this by attributing concurrent competences to different supervising authorities, by decentralizing enforcement in the presence of insufficient flows of information between the authorities, and by predominantly relying on public enforcement.

The last issue to be discussed in Chapter 6, and which ETS designers should take into account, involves criminal activities and the 'creative practices' of other ETS partner countries. In the context of the EU ETS, allowances were stolen and not all could be retrieved. The national allowance registries were not sufficiently protected and trading on the entire EU ETS market had to be suspended until improvements had been made. It is not only theft that can be a problem, but also fraud. If an emissions trading system incorporates various jurisdictions, fraudsters could collect VAT from purchasers of allowances in other states and fail to submit it to the government. Or they could claim reimbursement for the VAT allegedly paid on the allowances when, in fact, it has not been paid. ETS designers must take into account differences in VAT rules when establishing a multi-jurisdictional emissions trading system, or a system that allows cross-border trade in emission allowances. Last but not least, the case of Hungary was discussed. Hungary, while acting within the applicable legal regime, sold emission allowances (that it received for compliance purposes) abroad. Risks to consumers arose from the fact that allowances could be re-entering the EU ETS, since they could not have been used again for compliance purposes within the EU scheme. ETS designers should therefore ensure that such legal loopholes are closed in order to avoid instances that could undermine the credibility and the economic effectiveness of their emissions trading system. It is the task of the designer to shape the legal framework that allows the allowance market to prosper and enables it to function in a cost-effective way.

Chapter 7 examined issues relating to the operation of the system. It pointed towards the importance of good monitoring, reporting and verification schemes. If self-monitoring is relied upon, it must be inspected by competent authorities. The efficient working of such systems and the integrity of the individuals involved should not be taken for granted. ETS designers should also be very much aware that the benefits associated with emissions trading systems stem from cost-effective abatement by emitters that enjoy the lowest abatement costs.

Covered entities that are unable to take their abatement decisions on the basis of their abatement costs but are subject to administrative guidance or are directly owned by the state, may not be able to take the right decisions. It should therefore be ensured that the decisions of these entities will not be unduly influenced; if this cannot be ensured, policy designers should perhaps reconsider using emissions trading for those affected entities.

A rather technical issue that was addressed in the context of operational issues is in respect of the design of emission allowance transaction logs. These are databases that trace every transaction and hold information on the emission permits of the respective enterprises. It goes without saying that they should be allowed to start to operate only when they have been tested and are fully operational and capable of linking with other transaction logs. In the context of the EU they have been subject to an increasing level of privacy and data protection. Privacy and data protection standards must be maintained when an emissions trading system is to be linked with a foreign transaction log system. This can be difficult since different societies have developed their own standards of protection. High protection standards make it more difficult for researchers to provide policy makers with timely information on the actual operation of the emissions trading system. ETS designers will therefore need to make trade-offs between the benefits of both considerations.

Chapter 8 addressed some legal cases that have arisen in the context of climate change. They emanate from various jurisdictions, different emissions trading systems and different legal systems; therefore, they do not give rise to concrete legal insights that are directly applicable to other emissions trading systems. They do, however, provide an important insight for ETS designers – namely, that law matters. Emissions trading is a market-based instrument. This does not mean that it is a system designed by economists, is then checked by engineers for the technical details, and then simply implemented by lawyers. As is illustrated by the diversity of the cases reviewed in this chapter, there are a large number of legal issues that can arise in the context of emissions trading design, and they sometimes come from unexpected directions. They need to be addressed at the ETS design stage in order to prevent severe unforeseen circumstances that could grind the entire system or its design process to a halt. Legal insights, therefore, have an important part to play in both the design of the emissions trading system and its operation; their role should not be neglected by ETS designers.

5. WITH WHOM?

The last question addressed by this book is with whom an emissions trading system should be established. There is a rich and evolving body of literature examining which emissions trading design features allows for a beneficial link between different trading systems. Linking is attractive as it has the following three benefits: (i) linking of schemes can constitute an important 'second-best' solution to combat climate change in the event that international agreement is not forthcoming; (ii) from successful cooperation between nations a global solution for climate change could emerge; (iii) the linking of emissions trading systems allows for lower abatement costs.

From an ETS design perspective there are a number of insights to be observed. Linking with a larger scheme may involve the proliferation of the larger system's design features. When designing an emissions trading system, designers are therefore well advised to consider potential linking partners at the design stage in order to avoid costly adaptations to their system. Furthermore, ETS design features are not the only elements that deserve due attention: practical rules relating to, for example, monitoring, reporting and verification or the transaction log must also be considered. Incompatibility in those areas will require costly adaptations and will not create a level playing field for competing covered enterprises.

6. FINAL REMARKS

This book has attempted to take a more problem-oriented approach to emissions trading design. This stems from the perceived gap between the needs of policy designers, who normally attempt to realize multiple policy objectives on the one hand, and the topics perceived as 'interesting' by academics, on the other. This work is an attempt to assist policy makers in clarifying their policy objectives and design questions so as to successfully design an emissions trading system. Despite its undoubtedly existing shortcomings, it is hoped that it will help to build bridges between academia and societal needs so as to provide answers to social problems that are more readily associated with concrete policy problems. It is hoped that more research in the area of climate change will focus on conflicting policy objectives so that the works created in the remote ivory towers of academia become more meaningful to policymakers, and society will enjoy the benefits of better informed decision-making.

References

Adams, M. (1989), 'New activities and the efficient liability rules', in M. Faure and R. Van den Bergh (eds), *Essays in Law and Economics. Corporations, Accident Prevention and Compensation for Losses*, Antwerp: Maklu uitgevers.

Aggarwal, R.K. and G. Wu (2006), 'Stock market manipulations', *The Journal of Business*, **79** (4), 1915–53.

Aldy, J.E. and R.N. Stavins (2011), 'The promise and problems of pricing carbon: theory and experience', Harvard Institute of Economic Research Discussion Paper, available at: http://papers.ssrn.com/sol3/papers.cfm?abstract_id=1950693 (last accessed 15 September 2013).

Arcuri, A. (2001), 'Controlling environmental risk in Europe: the complementary role of an EC environmental liability regime', *Tijdschrift voor Milieuaansprakelijkheid* (Journal of Environmental Liability), 39–40.

Ashenfelter, O. (1989), 'How auctions work for wine and art', *Journal of Economic Perspectives*, **3** (3), 23–36.

Atsma, J. (2012), Letter from Joop Atsma (Secretary of State for Infrastructure and the Environment) to the Chairman of the Lower Chamber, 4 July 2012, on 'Brief met toezeggingen n.a.v. het AO klimaatbeleid d.d. 31 mei 2012' ('Letter with commitments based upon the general discussions on climate policy dated 31 May 2012').

Auction Regulation: Commission Regulation (EU) No. 1031/2010 of 12 November 2010 on the timing, administration and other aspects of auctioning of greenhouse gas emission allowances pursuant to Directive 2003/87/EC of the European Parliament and of the Council establishing a scheme for greenhouse gas emission allowances trading within the Community, OJ L 302/1, 18 November 2010.

Ausubel, L. and P. Cramton (2002), *Demand Reduction and Inefficiency in Multi-Unit Auctions*, University of Maryland.

Ballard, C., J. Shoven and J. Whalley (1985), 'General equilibrium computations of the marginal welfare costs of taxes in the United States', *American Economic Review*, **75**, 128–38.

Baron, R. and Philibert, C. (2005), 'Act locally, trade globally: emissions trading for climate policy', International Energy Agency, available at: http://www.iea.org/publications/freepublications/publication/act_locally.pdf

Baumol, W.J. and W.E. Oates (1988), *The Theory of Environmental Policy*, 2nd edn, Cambridge, UK: Cambridge University Press.

Bazelmans, J. (2008), 'Linking the EU ETS to other emissions trading schemes', in M. Faure and M. Peeters (eds), *Climate Change and European Emissions Trading: Lessons for Theory and Practice*, Cheltenham, UK and Northampton, MA, USA: Edward Elgar, 297–321.

Behr, T., J.M. Witte, W. Hoxtell and J. Manzer (2009), 'Towards a global carbon market? Potential and limits of carbon market integration', Energy Policy Paper Series, Global Public Policy Institute, Berlin, available at: http://www.gppi.net/fileadmin/gppi/GPPiPP7-Carbon_Markets.pdf.

Bergkamp, L. (2001), *Liability and Environment*, The Hague and London: Kluwer Law International.

Bocken, H. (1987), 'Alternatives to liability and liability insurance for the compensation of pollution damages', *Tijdschrift voor Milieuaansprakelijkheid* (Journal of Environmental Liability), 83–87.

Bocken, H. (1988), 'Alternatives to liability and liability insurance for the compensation of pollution damages', *Tijdschrift voor Milieuaansprakelijkheid* (Journal of Environmental Liability), 3–10.

Bohm, P. (1999), *International Greenhouse Gas Emission Trading – With Special Reference to the Kyoto Protocol*, TemaNord 1999:506, Stockholm: Department of Economics.

Böhringer, C., T. Hoffmann, A. Lange, A. Löschel and U. Moslener (2005), 'Assessing emission regulation in Europe: an interactive simulation approach', *Energy Journal*, **26**, 1–22.

Bovenberg, A. and L. Goulder (2000), 'Neutralizing the adverse industry impacts of CO_2 abatement policies: What does it cost?', NEBR Working Paper Series, No. 7654, 34.

Brander, L. (2003), 'The Kyoto mechanisms and the economics of their design', in M. Faure, J. Gupta and A. Nentjes (eds), *Climate Change and the Kyoto Protocol: The Role of Institutions and Instruments to Control Global Change*, Cheltenham, UK and Northampton, MA, USA: Edward Elgar, 25–44.

Brown, J.P. (1973), 'Toward an economic theory of liability', *Journal of Legal Studies* **2** (2), 323–49.

Burrows, P. (1999), 'Combining regulation and liability for the control of external costs', *International Review of Law and Economics*, **19**, 227–42.

Calabresi, G. (1961), 'Some thoughts on risk distribution and the law of torts', *Yale Law Journal*, **70**, 499–553.

Calabresi, G. (1970), *The Costs of Accidents. A Legal and Economic Analysis*, New Haven, CT, USA: Yale University Press.

Calabresi, G. (1975), 'Optimal deterrence and accidents', *Yale Law Journal*, **84**, 656–71.

California Air Resources Board (2012), 'California cap on greenhouse gas emissions and market-based compliance mechanisms', 1 September 2012, available at: http://www.arb.ca.gov/cc/capandtrade/september_2012_regulation.pdf (last accessed 15 September 2013).

California Air Resources Board (2013), 'Air Resources Board sets date for linking cap-and-trade program with Québec', Press Release, 19 April 2013, available at: http://www.arb.ca.gov/newsrel/newsrelease.php?id=430 (last accessed 15 September 2013).

Cao, M. (2011), 'China's law development in the climate change era', available at: http://ssrn.com/abstract=1832006 (last accessed 15 September 2013).

Carbon Market News (2013), 'Carbon Market North America January 11', 11 January 2013, available at: http://www.pointcarbon.com/news/cmna/1.2134387.

Chalmers, D. and A. Tomkins (2007), *European Union Public Law, Cases and Materials*, Cambridge, UK: Cambridge University Press.

Charter of Fundamental Rights of the European Union (2000), OJ C 364/01, 18 December 2000.

'Chongqing will launch carbon trading, high energy consumption industries are involved', available at: http://www.coal.com.cn/Gratis/2012-11-21/ArticleDisplay_329613.shtml. For future information: http://www.cquae.com/html/tpfjy.

Clean Energy Amendment (International Emissions Trading and Other Measures) Bill 2012, 26 November 2012, available at: http://www.aph.gov.au/Parliamentary_Business/Bills_Legislation/Bills_Search_Results/Result?bId=r4895.

Clò, S. (2007), 'Assessing the European emissions trading scheme effectiveness in reaching the Kyoto Target: an analysis of the ETS 1st and 2nd phase cap stringency', Working Paper presented at the Annual Conference of the European Association of Law and Economics 2007, available at: http://www.cbs.dk/content/download/67304/930289/file/Stefano%20Clň.pdf.

CO_2 Gesetz: Bundesgesetz über die Reduktion der CO_2-Emissionen, 641.71vom 23. Dezember 2011 (Stand am 1. Januar 2013) available at: http://www.admin.ch/ch/d/sr/6/641.71.de.pdf

CO_2 Verordnung: Verordnung über die Reduktion der CO_2-Emissionen, 641.711vom 30. November 2012 (Stand am 1. Juni 2013) available at: http://www.admin.ch/ch/d/sr/6/641.711.de.pdf.

COM (2008) 30 final, EC Commission Communication, '20 20 by 2020: Europe's climate change opportunity', Brussels, 23 January 2008.

COM (2010) 796 final, European Commission Communication, 'Towards an enhanced market oversight framework for the EU emissions trading scheme', Brussels, 21 December 2010.
COM (2010) 2020 final, European Commission Communication, 'Europe 2020: a strategy for smart, sustainable and inclusive growth', Brussels, 3 March 2010.
COM (2011) 109 final, European Commission Communication, 'Energy Efficiency Plan 2011', Brussels, 5 May 2011.
COM (2011) 370 final, European Commission, 'Proposal for a Directive of the European Parliament and of the Council on energy efficiency and repealing Directives 2004/8/EC and 2006/32/EC', Brussels, 22 June 2011.
COM (2011) 656 final, 2011/0298 (COD), European Commission, 'Proposal for a Directive of the European Parliament and of the Council on markets in financial instruments repealing Directive 2004/39/EC of the European Parliament and of the Council (Recast)', Brussels, 20 October 2011.
COM (2012) 11 final, 2012/0011 (COD), European Commission, 'Proposal for a Regulation of the European Parliament and of the Council on the protection of individuals with regard to the processing of personal data and on the free movement of such data (General Data Protection Regulation)', Brussels, 25 January 2012.
COM (2012) 416 final, 2012/0202 (COD), 'Proposal for a decision of the European Parliament and of the Council amending Directive 2003/87/EC clarifying provisions on the timing of auctions of greenhouse gas allowances', Brussels, 25 July 2012.
COM (2012) 697, 2012/328 (COD), European Commission, 'Proposal for a Decision of the European Parliament and of the Council derogating temporarily from Directive 2003/87/EC of the European Parliament and of the Council establishing a scheme for greenhouse gas emission allowance trading within the Community', Strasbourg.
Commission Decision 2011/278/EU of 27 April 2011 determining transitional Union-wide rules for harmonised free allocation of emission allowances pursuant to Article 10a of Directive 2003/87/EC of the European Parliament and of the Council (notified under document C(2011) 2772), OJ L 130/1, 17 May 2011.
Commission Regulation (EU) No. 920/2010 of 7 October 2010 for a standardised and secured system of registries pursuant to Directive 2003/87/EC of the European Parliament and of the Council and Decision No 280/2004/EC of the European Parliament and of the Council, OJ L 270/1, 14 October 2010.

Commission Regulation (EU) No. 1031/2010 of 12 November 2010 on the timing, administration and other aspects of auctioning of greenhouse gas emission allowances pursuant to Directive 2003/87/EC of the European Parliament and of the Council establishing a scheme for greenhouse gas emission allowances trading within the Community, OJ L 302/1, 18 November 2010 (Auction Regulation).

Commission Regulation (EU) No. 550/2011 of 7 June 2011 on determining, pursuant to Directive 2003/87/EC of the European Parliament and of the Council, certain restrictions applicable to the use of international credits from projects involving industrial gases, OJ L 149/1, 8 June 2011.

Commission Regulation (EU) No. 1193/2011 of 18 November 2011 establishing a Union Registry for the trading period commencing on 1 January 2013, and subsequent trading periods, of the Union emissions trading scheme pursuant to Directive 2003/87/EC of the European Parliament and of the Council and Decision No. 280/2004/EC of the European Parliament and of the Council and amending Commission Regulations (EC) No. 2216/2004 and (EU) No. 920/2010, OJ L 315/1, 29 November 2011.

Commission Regulation (EU) No. 600/2012 of 21 June 2012 on the verification of greenhouse gas emission reports and tonne-kilometre reports and the accreditation of verifiers pursuant to Directive 2003/87/EC of the European Parliament and of the Council, OJ L 181/1, 12 July 2012.

Commission Regulation (EU) No. 601/2012 of 21 June 2012 on the monitoring and reporting of greenhouse gas emissions pursuant to Directive 2003/87/EC of the European Parliament and of the Council, OJ L 181/30, 12 July 2012.

Commission Regulation (EU) No. 389/2013 of 2 May 2013 establishing a Union Registry pursuant to Directive 2003/87/EC of the European Parliament and of the Council, Decisions No. 280/2004/EC and No. 406/2009/EC of the European Parliament and of the Council and repealing Commission Regulations (EU) No. 920/2010 and No. 1193/2011, OJ L 122/1, 3 May 2013.

Commission Staff Working Document (2012), 'Information provided on the functioning of the EU Emissions Trading System, the volumes of greenhouse gas emission allowances auctioned and freely allocated and the impact on the surplus of allowances in the period up to 2020', Brussels, 25 July 2012, SWD(2012) final, available at: http://ec.europa.eu/clima/policies/ets/cap/auctioning/docs/swd_2012_234_en.pdf (last accessed 24 September 2013).

Commonwealth of Australia (2011), 'Securing a clean energy future: the Australian Government's climate change plan', available at: http://www.

cleanenergyfuture.gov.au/wp-content/uploads/2011/07/Consolidated-Final.pdf (last accessed 4 August 2011), currently available at: http://large.stanford.edu/courses/2012/ph240/aslani2/docs/CleanEnergyPlan-20120628-3.pdf.

Communication from the Commission (2012), Guidelines on certain State aid measures in the context of the greenhouse gas emission allowance trading scheme post-2012, SWD(2012) 130 final, OJ C 158/4, 5 June 2012.

Cooter, R. (1984), 'Prices and sanctions', *Columbia Law Review*, **84**, 1343–523.

Copenhagen Accord (2009) FCCC/CP/2009/11/Add.1, 30 March 2010.

Costa, L. and Y. Poullet (2012), 'Privacy and the regulation of 2012', *Computer Law & Security Review*, **28**, 254–62.

Council of the European Union (2009), 'Council adopts climate-energy legislative package', 8434/09 (Presse 77), Brussels, 6 April 2009, available at: http://europa.eu/rapid/press-release_PRES-09-77_en.htm.

Council Directive 2010/23/EU of 16 March 2010 amending Directive 2006/112/EC on the common system of value added tax, as regards an optional and temporary application of the reverse charge mechanism in relation to supplies of certain services susceptible to fraud, OJ L 72/1, 20 March 2010.

Council Directive 2013/43/EU of 22 July 2013 amending Directive 2006/112/EC on the common system of value added tax, as regards an optional and temporary application of the reverse charge mechanism in relation to supplies of certain goods and services susceptible to fraud, OJ L 201/4, 26 July 2013.

Cramton, P. and S. Kerr (1999), 'Tradable carbon permit auctions: how and why to auction, not grandfather', Wharton, Financial Institutions Center, Implications for Auction Theory for New Issues Markets, No. 02-19.

Criqui, P. and A. Kitous (2003), 'Kyoto Protocol implementation – KPI Technical Report: Impacts of linking JI and CDM credits to the European Emission Allowance Trading Scheme – a Report for DG Environment', May 2003, CNRS-IEPE and Enerdata SA.

Cullis, J. and P. Jones (1998), *Public Finance and Public Choice*, 2nd edn, Oxford: Oxford University Press.

Dales, J. (1968), *Pollution, Property and Prices: An Essay in Policy*, Toronto: University of Toronto Press.

De Cendra de Larragán, J. (2006), 'Linking the project based mechanism with the EU ETS: the present state of affairs and challenges ahead', in M. Peeters and K. Deketelaere (eds), *EU Climate Change Policy: The Challenge of New Regulatory Initiatives*, Cheltenham, UK and Northampton, MA, USA: Edward Elgar, 98–124.

De Cendra de Larragán, J. (2013), 'Linking planetary boundaries and overconsumption by individuals', in S. Kingston (ed), *European Perspectives on Environmental Law and Governance*, Abingdon, UK: Routledge, 23–54.

De Hert, P. and S. Gutwirth (2009), 'Data protection in the case law of Strasbourg and Luxembourg: Constitutionalisation in action', in S. Gutwirth et al. (eds), *Reinventing Data Protection?*, Dordrecht: Springer Science, 3–44.

De Mooij, R.A. (1999), 'The double dividend of an environmental tax reform', in J.C.J.M. van den Bergh (ed), *Handbook of Environmental and Resource Economics*, Cheltenham, UK and Northampton, MA, USA: Edward Elgar, 293–306.

Decision 13/CMP.1, Modalities for the accounting of assigned amounts under Article 7, paragraph 4 of the Kyoto Protocol, available at: http://unfccc.int/resource/docs/2005/cmp1/eng/08a02.pdf.

Decision No. 406/2009/EC of the European Parliament and of the Council of 23 April 2009 on the effort of Member States to reduce their greenhouse gas emissions to meet the Community's greenhouse gas emission reduction commitments up to 2020, OJ L 140, 5 June 2009.

Department of Climate Change and Energy Efficiency (2011), 'Exposure draft of the Clean Energy Bill 2011: commentary on provisions', 28 July 2011, available at: http://apo.org.au/sites/default/files/Commentary-on-the-Clean-Energy-Bill-2011-PDF.pdf.

Diamond, P. (1974), 'Single activity accidents', *Journal of Legal Studies*, **3**, 107–64.

Diaz-Rainey, I., M. Siems and J.K. Ashton (2011), 'The financial regulation of European wholesale energy and environmental markets', USAEE – IAEE Working Paper 11-070.

Dijkstra, B.R. (1999), *The Political Economy of Environmental Policy: A Public Choice Approach to Market Instruments*, Cheltenham, UK and Northampton, MA, USA: Edward Elgar.

Directive 2003/6/EC of the European Parliament and of the Council of 28 January 2003 on insider dealing and market manipulation (market abuse) OJ L 96/16, 12 April 2003.

Directive 2003/87/EC of the European Parliament and of the Council of 13 October 2003 establishing a scheme for greenhouse gas emission allowance trading within the Community and amending Council Directive 96/61/EC, OJ L 275/32, 25 October 2003.

Directive 2004/39/EC of the European Parliament and of the Council of 21 April 2004 on markets in financial instruments amending Council Directives 85/611/EEC and 93/6/EEC and Directive 2000/12/EC of the

European Parliament and of the Council and repealing Council Directive 93/22/EEC, OJ L 145/1, 30 April 2004.
Directive 2004/101/EC of the European Parliament and of the Council of 27 October 2004 amending Directive 2003/87/EC establishing a scheme for greenhouse gas emission allowance trading within the Community, in respect of the Kyoto Protocol's project mechanisms, OJ L 338/18, 13 November 2004.
Directive 2005/60/EC on the prevention of the use of the financial system for the purpose of money laundering and terrorist financing, OJ L 309/15, 25 November 2005.
Directive 2009/29/EC of the European Parliament and of the Council of 23 April 2009 amending Directive 2003/87/EC so as to improve and extend the greenhouse gas emission allowance trading scheme of the Community, OJ L 140/63, 5 June 2009.
Draft Commission Regulation amending Regulation (EU) No. 1031/2010 in particular to determine the volumes of greenhouse gas emission allowances to be auctioned in 2013–2020, available at: http://ec.europa.eu/clima/policies/ets/reform/docs/2013_07_08_en.pdf.
ECN & Cambridge Econometrics (2013), 'Splitting the EU ETS: strengthening the scheme by differentiating its sectoral carbon prices', May 2013, ECN-E-13-008, available at: http://www.camecon.com/Libraries/Downloadable_Files/Improving_the_EU_ETS__splitting_the_power_and_industrial_sectors_report.sflb.ashx.
EDPS Opinion (2012): Opinion of the European Data Protection Supervisor of 11 May 2012 on the Commission Regulation establishing a Union Registry for the trading period commencing on 1 January 2013, and subsequent trading periods, of the Union emissions trading scheme, available at: http://www.edps.europa.eu/EDPSWEB/webdav/site/mySite/shared/Documents/Consultation/Opinions/2012/12-05-11_Trading_period_EN.pdf (last accessed 15 May 2013).
EEX (2011), 'EEX Product Brochure, EU Emission Allowances', available at: http://cdn.eex.com/document/89518/20110329_EEX_Produktbrosch%C3%BCre_CO2_engl.pdf.
Egenhofer, C. and N. Fujiwara (2005), *Reviewing the EU Emissions Trading Scheme: Priorities for Short-Term Implementation of the Second Round of Allocation (Part I)*, Brussels: Centre for European Policy Studies (CEPS).
Ellerman, A.D. (2004), 'The US SO_2 Cap-and-trade program', in Organisation for Economic Co-operation and Development, *Tradable Permits, Policy Evaluation, Design and Reform*, Paris: OECD.

Ellerman, D. and B. Buchner (2006), *Over-Allocation or Abatement? A Preliminary Analysis of the EU ETS Based on the 2005 Emissions Data*, Nota di Lavoro 139.2006, Milan: Fondazione Eni Enrico Mattei (FEEM).

Endres, A. and R. Schwarze (1991), 'Allokationswirkungen einer Umwelthaftpflichtversicherung' (Allocative effects of environmental liability insurance), *Zeitschrift fuer Umweltpolitik und Umweltrecht* (*Journal for Environmental Policies and Environmental Law*), **14**, 1–25.

Environmental Quality Act, Regulation respecting a cap-and-trade system for greenhouse gas emission allowances, Quebec, Chapter Q-2, r. 46.1 (Chapter Q-2, s. 31, 1st para., subparas b, c, d, e.1, h and h.1, ss. 46.1, 46.5, 46.6, 46.8 to 46.16, 115.27 and 115.34), as of 1 November 2013, available at: http://www2.publicationsduquebec.gouv.qc.ca/dynamicSearch/telecharge.php?type=3&file=/Q_2/Q2R46_1_A.HTM.

Eurometaux (2011), 'Guidelines on certain state aid measures in the context of the Greenhouse Gas Emission Allowance Trading Scheme: response to the DG Competition Consultation as published on their website on 21 December 2011', available at: http://ec.europa.eu/competition/consultations/2012_emissions_trading/eurometaux_en.pdf.

European Commission (2003), 'Systeem van verhandelbare emissierechten voor NOx', C(2003)1761fin, 24 June 2003.

European Commission (2005), 'EU emissions trading: an open scheme promoting global innovation to combat climate change', available at: http://www.pedz.uni-mannheim.de/daten/edz-bn/gdu/05/emission_trading2_en.pdf.

European Commission (2011), 'Guidance on new entrants and closures', Guidance Document No. 7 on the Harmonized Free Allocation Methodology for the EU-ETS Post 2012, 14 September 2011.

European Commission (2012a), 'Renewables: Commission confirms market integration and the need for growth beyond 2020', Press Release, IP/12/571, 6 June 2012, available at: http://europa.eu/rapid/press-release_IP-12-571_en.htm.

European Commission (2012b), 'Stopping the clock of ETS and aviation emissions following last week's International Civil Aviation Organization (ICAO) Council', MEMO 12/854, Brussels, 12 November 2012, available at: http://europa.eu/rapid/press-release_MEMO-12-854_en.htm.

European Commission (2012c), 'Commission submits formal proposal to defer EU ETS international aviation compliance by one year', 20 November 2012, available at: http://ec.europa.eu/clima/news/articles/news_2012112001_en.htm.

European Commission (2012d), 'Europe 2020 Targets: Climate Change and Energy', available at: http://ec.europa.eu/europe2020/pdf/themes/16_energy_and_ghg.pdf.

European Commission (2013), Commission Regulation on determining international credit entitlements pursuant to Directives 2003/87/EC of the European Parliament and of the Council of 5 June 2013, available at: http://ec.europa.eu/clima/policies/ets/docs/c_2013_xxx_en.pdf (last accessed 13 June 2013).

European Convention for the Protection of Human Rights and Fundamental Freedoms, entered into force 3 September 1953, 213 UNTS 222.

European Union Emissions Trading Scheme Prohibition Act of 2011 (Enrolled Bill) [Final as Passed Both House and Senate], S. 1956, available at: http://www.gpo.gov/fdsys/pkg/BILLS-112s1956enr/pdf/BILLS-112s1956enr.pdf.

Fankhauser, S. and C. Hepburn (2010), 'Designing carbon markets – Part I: carbon markets in time', *Energy Policy*, **38**, 4363–70.

Faure, M. (1995), 'Economic models of compensation for damage caused by nuclear accidents: some lessons for the revision of the Paris and Vienna Conventions', *European Journal of Law and Economics*, **2** (1), 21–43.

Faure, M. (2012), 'Effectiveness of environmental law: What does the evidence tell us?', *William & Mary Environmental Law and Policy Review*, **36**, 293–336.

Faure, M. and M. Peeters (2011), *Climate Change Liability*, Cheltenham, UK and Northampton, MA, USA: Edward Elgar.

Faure, M., M. Peeters and A. Wibisana (2006), 'Economic instruments: suited to developing countries?', in M. Faure and N. Niessen (eds), *Environmental Law in Development: Lessons from the Indonesian Experience*, Cheltenham, UK and Northampton, MA, USA: Edward Elgar, 218–62.

Faure, M. and M. Ruegg (1994), 'Standard setting through general principles of environmental law', in M. Faure, J. Vervaele and A. Weale (eds), *Environmental Standards in the European Union in an Interdisciplinary Framework*, Antwerpen-Apeldoorn, the Netherlands: Maklu uitgevers, 39–60.

Faure, M. and S. Ubachs (2003), 'Comparative benefits and optimal use of environmental taxes', in J. Milne et al. (eds), *Critical Issues in Environmental Taxation: International and Comparative Perspectives*, Vol. I, Richmond, UK: Richmond Law and Tax, 29–49.

Faure, M. and S.E. Weishaar (2012), 'The role of environmental taxation: economics and the law', in J. Milne and M.S. Andersen (eds),

Handbook of Research on Environmental Taxation, Cheltenham, UK and Northampton, MA, USA: Edward Elgar, 399–421.

Fawcett, T. and Y. Parag (2010), 'An introduction to personal carbon trading', *Climate Policy*, **10**, 329–38.

Feldman, Y. and O. Perez (2009), 'How law changes the environmental mind: an experimental study of the effect of legal norms on moral perceptions and civic enforcement', *Journal of Law and Society*, **36** (4), 501–35.

Fell, H., E. Moore and R.D. Morgenstern (2011), 'Cost containment under cap and trade: a review of the literature', *International Review of Environmental and Resource Economics*, **5**, 285–307.

Flachsland, C., R. Marschinski and O. Edenhofer (2009a), 'To link or not to link: benefits and disbenefits of linking cap-and-trade systems', *Climate Policy*, **9** (4) Special Issue *Linking GHG Trading Systems*, 358–72.

Flachsland, C., R. Marschinski and O. Edenhofer (2009b), 'Global trading versus linking: architectures for international emissions trading', *Energy Policy*, **37** (5), 1637–47.

Frank, R. (1997), *Microeconomics and Behaviour*, 3rd edn, Boston, MA, USA: Irwin/McGraw-Hill.

Fullerton, D., A. Leicester and S. Smith (2010), 'Environmental taxes', in Institute for Fiscal Studies (ed), *Dimensions of Tax Design*, Oxford: Oxford University Press, 423–547.

General Data Protection Regulation: COM(2012) 11 final, 2012/0011 (COD), European Commission, 'Proposal for a Regulation of the European Parliament and of the Council on the protection of individuals with regard to the processing of personal data and on the free movement of such data (General Data Protection Regulation)', Brussels, 25 January 2012.

Goulder, L.H. (1995), 'Environmental taxation and the "double dividend": a reader's guide', *International Tax and Public Finance*, **2** (2), 157–83.

Granaut, R. (2011), *The Granaut Climate Change Review: The Final Report*, Cambridge, UK: Cambridge University Press.

Groosman, B. (1999), '2500 pollution tax', in B. Bouckaert and G. De Geest, *Encyclopedia of Law and Economics: Volume II, Civil Law and Economics*, Cheltenham, UK and Northampton, MA, USA: Edward Elgar, 538–68.

Grull, G. and L. Taschini (2010), 'Linking emissions trading schemes: a short note', available at: http://ssrn.com/abstract=1546105.

Gullì, F. (ed) (2008), *Markets for Carbon and Power Pricing in Europe: Theoretical Issues and Empirical Analyses*, Cheltenham, UK and Northampton, MA, USA: Edward Elgar.

Haites, E. (2003), 'Harmonization between national and international tradeable permit schemes', CATEP Synthesis Paper, OECD, Paris.
Hansen, J. (2009), *Storms of My Grandchildren: The Truth About the Coming Climate Catastrophe and Our Last Chance to Save Humanity*, New York, NY, USA: Bloomsbury.
Harrison, D. and D. Radov (2002), 'Evaluation of alternative initial allocation mechanisms in a European Union Greenhouse Gas Emissions Allowance Trading Scheme', National Economic Research Associates, available at: http://www.nera.com/67_4804.htm.
Heilmayr, R. and J.A. Bradbury (2011), 'Effective, efficient or equitable: using allowance allocations to mitigate emissions leakage', *Climate Policy*, **11**, 1113–30.
Hepburn, C., M. Grubb, K. Neuhoff, F. Matthes and M. Tse (2006) 'Auctioning of EU ETS Phase II allowances: how and why?' *Climate Policy*, **6** (1), 137–60.
Himma, K.E. (2007), 'Privacy vs. security: why privacy is not an absolute value or right', *San Diego Law Review*, **45**, available at: http://ssrn.com/abstract=994458 (last accessed 13 February 2013).
'Hubei Provinces will officially launch carbon trading in August 2013', available at: http://www.tanpaifang.com/tanjiaoyi/2013/0419/19604.html.
'Hubei steadily explores the way for Middle and Western regions to promote carbon trading', available at: http://info.china.alibaba.com/detail/1123025658.html.
'Implementation plan for Guangdong Province's carbon trading', available at: http://zwgk.gd.gov.cn/006939748/201209/t20120914_343489.html. For future information: http://www.nea.gov.cn/2013-02/26/c_132192742.htm.
'Implementation plan of Shanghai's carbon trading', available at: http://www.shanghai.gov.cn/shanghai/node2314/node2319/node12344/u26ai32789.html. For future information, visit http://www.cneeex.com.
'Implementation plan of Tianjin's emission trading pilot', available at: http://www.tjzfxxgk.gov.cn/tjep/ConInfoParticular.jsp?id=38237. For future information: http://www.chinatcx.com.cn/tcxweb.
Intergovernmental Panel on Climate Change (IPCC) (2007), 'Fourth Assessment Report (2007)', available at: http://www.ipcc.ch/publications_and_data/publications_ipcc_fourth_assessment_report_synthesis_report.htm.
International Covenant on Civil and Political Rights, adopted by UNGA Resolution 2200 (XXI) of 16 December 1966, entered into force 23 March 1976, 999 UNTS 171.
Jaffe, J. and R.N. Stavins (2007), 'Linking tradable permit systems for greenhouse gas emissions: opportunities, implications and challenges',

International Emissions Trading Association, Geneva, available at: http://belfercenter.ksg.harvard.edu/files/IETA_Linking_Report.pdf.
Jakob-Gallman, J. (2011), *Regulatory Issues in the Carbon Market: The Linkage of the Emissions Trading Scheme of Switzerland with the Emissions Trading Scheme of the European Union*, Zurich: Schulthess.
Jepma, C. (2006), 'Some EU ETS "tags"', *(Electronic) Joint Implementation Quarterly: Magazine on the Kyoto Mechanisms*, August 2006, 1–7.
Jones, M.C. et al. (2011), 'Australia – finally a carbon pricing scheme?', in J. Peetermans (ed), *Greenhouse Gas Market Report 2011 – Asia and Beyond: the Roadmap to Global Carbon & Energy Markets*, Geneva: International Emissions Trading Association, available at http://www.ieta.org/assets/Reports/ieta%202011%20ghg%20market%20report%20final.pdf.
Jong, M.A.P., O. Couwenberg and E. Woerdman (2013), 'Does the EU ETS bite? The impact of allowance over-allocation on share prices', University of Groningen: Working Paper Series in Law and Economics.
Jordan, B.D. and S.D. Jordan (1996), 'Salomon Brothers and the May 1991 Treasury auction: analysis of a market corner', *Journal of Banking & Finance*, **20**, 25–40.
Jotzo, F. (2013), 'Emissions trading in China: principles, design options and lessons from international practice', Centre for Climate Economics & Policy, CCEP Working Paper 1303, May 2013.
Kaminskaitė-Salters, G. (2011), 'Climate change litigation in the UK: its feasibility and prospects', in M. Faure and M. Peeters, *Climate Change Liability*, Cheltenham, UK and Northampton, MA, USA: Edward Elgar, 165–88.
Kettner, C., A. Köppl, S.P. Schleicher and G. Thenius (2007), 'Stringency and distribution in the EU emissions trading scheme – the 2005 evidence', FEEM Working Paper 22.2007.
Kilkelly, U. (2001), 'The Right to Respect for Private and Family Life', Human Rights Handbooks, No. 1, available online at: http://www.echr.coe.int/library/DIGDOC/DG2/HRHAND/DG2-EN-HRHAND-01(2003).pdf (last accessed 16 September 2013).
Kleining, P. et al. (2011), *Security and Privacy: Global Standards for Ethical Identity Management in Contemporary Liberal Democratic States*, Canberra, ACT: ANU E Press.
Knudson, W.A. (2009), 'The environment, energy and the Tinbergen Rule', *Bulletin of Science Technology & Society*, **29**, 308–12.
Kolstad, C.D., T.S. Ulen and G.V. Johnson (1990), 'Ex post liability for harm vs. ex ante safety regulation: substitutes or complements?', *American Economics Review*, **80**, 888–901.

Kosolapova, E. (2011), 'Liability for climate change-related damage in domestic courts: claims for compensation in the USA', in M. Faure and M. Peeters (eds), *Climate Change Liability*, Cheltenham, UK and Northampton, MA, USA: Edward Elgar, 189–205.

Krishna, V. (2002), *Auction Theory*, San Diego: Academic Press.

Kunreuther, H. and P. Freeman (2001), 'Insurability, environmental risks and the law', in A. Heyes (ed), *The Law and Economics of the Environment*, Cheltenham, UK and Northampton, MA, USA: Edward Elgar, 304–05.

Kyoto Protocol to the United Nations Framework Convention on Climate Change, United Nations, 1998, available at http://unfccc.int/resource/docs/convkp/kpeng.pdf.

Landes, W. and R. Posner (1984), 'Tort law as a regulatory regime for catastrophic personal injuries', *Journal of Legal Studies*, **13**, 417–34.

Lauge Pedersen, S. (2000), 'The Danish CO_2 emissions trading system', *Review of European Community & International Environmental Law*, **9** (3), 223–31.

Lecourt, S., C. Pallière and O. Sartor (2013), 'Free allocations in EU ETS Phase 3: the impact of emissions-performance benchmarking for carbon-intensive industry', Les Cahiers de la Chaire Economie du Climat Working Paper Series 2013-02, Dauphine University, Paris.

Lederer, M. (2012), 'Market making via regulation: the role of the state in carbon markets', *Regulation & Governance*, 1–21.

Leinweber, D.J. and A.N. Madhaven (2001), 'Three hundred years of stock market manipulations', *Journal of Investing*, 1–10.

Lieb, C.M. (2004), 'The environmental Kuznets curve and flow versus stock pollution: the neglect of future damages', *Environmental & Resource Economics*, **29** (4), 483–506.

Mackaay, E. (1982), *Economics of Information and the Law*, Boston, MA, USA: Kluwer Nijhoff.

Marrakech Accords (2001), Marrakesh Accords and the Marrakesh Declaration, adopted by the Conference of the Parties at its seventh session, November 2001, unedited form, available at: http://unfccc.int/cop7/documents/accords_draft.pdf.

Marx, G.T. (1998), 'Ethics for the New Surveillance', *The Information Society*, **14** (3), 171–85.

Mattoo, A., A. Subramanian, D. van der Mensbrugghe and J. He (2009), 'Reconciling climate change and trade policy', World Bank Policy Research, Working Paper 5123, available at: http://elibrary.worldbank.org/content/workingpaper/10.1596/1813-9450-5123.

McKibbin, W.J., and P.J. Wilcoxen (2002), 'The role of economics in climate change policy', *Journal of Economic Perspectives*, **16** (2), 107–29.

Milaj, J. (2013), 'Incidentally intercepted communications – a challenge to privacy?', Paper presented at the 'Online Privacy: Consenting to your Future' Conference in Malta, 20–21 March 2013. Paper available from the author.

Milgrom, P. (2004), *Putting Auction Theory to Work*, Cambridge, UK: Cambridge University Press.

NAP-NL (2007), Nederlands nationaal toewijzingsplan broeikasgasemissierechten 2008–2012, Plan van de Minister van Economische Zaken en de Minister van Volkshuisvesting, Ruimtelijke Ordening en Milieubeheer, 16 mei 2007, Consolidated version (Stcrt. 2008, nr. 132) (in Dutch), available at: https://www.emissieautoriteit.nl/mediatheek/emissierechten/toewijzing-emissierechten/toewijzingsbesluiten/NAP2%20def%20goedgekeurd_tcm24-282985.pdf.

NDRC (2010), 'The notice to pilot provincial and city low-carbon programmes', No. 1587 (in Sandbag (2012)).

NDRC (2012), 'China's policies and actions for addressing climate change', available at: http://www.ccchina.gov.cn/WebSite/CCChina/UpFile/File1324.pdf.

New Zealand Government – Climate Change Information (2013), 'Legislative changes to the New Zealand Emissions Trading Scheme (NZ ETS)', available at: http://climatechange.govt.nz/emissions-trading-scheme/ets-amendments (last accessed 17 May 2013)

Nield, K. and R. Pereira (2011), 'Fraud on the European Union Emissions Trading Scheme: effects, vulnerabilities and regulatory reform', *European Energy and Environmental Law Review*, **20** (6), 255–89.

Nishida, Y. and Y. Jua (2011), 'Motivating stakeholders to deliver change: Tokyo's cap-and-trade program', *Building Research & Innovation*, **39** (5), 518–33.

OECD (1999), *Economic Instruments for Pollution Control and Natural Resources Management in OECD Countries: A Survey*, Paris: Organisation for Economic Co-operation and Development.

Office of the Governor of California, Executive Order S-3-05 of 6 January 2005, available at: http://www.dot.ca.gov/hq/energy/ExecOrderS-3-05.htm.

Order of the Council 1187-2009 of 19 November 2009, Decree 1187-2009 on the adoption of Quebec's GHG target by 2020, *Gazette Officielle du Québec*, **2** (49), 9 December 2009, p. 5871, available in French at: http://www2.publicationsduquebec.gouv.qc.ca/dynamicSearch/telecharge.php?type=1&file=52750.PDF.

Order of the Council 1185-2012 of 12 December 2012, Determination of annual caps on greenhouse gas emission units relating to the cap-and-trade system for greenhouse gas emission allowances for the 2013–2020 period, *Gazette Officielle du Québec*, 19 December 2012,

144 (51), available at: http://www2.publicationsduquebec.gouv.qc.ca/dynamicSearch/telecharge.php?type=1&file=2389.PDF.

Parliamentary Amendment Bill of 26 November 2012, 'Clean Energy Amendment (International Emissions Trading and Other Measures)', available at: http://parlinfo.aph.gov.au/parlInfo/download/legislation/bills/r4895_aspassed/toc_pdf/12167b01.pdf;fileType=application%2Fpdf (last accessed 29 April 2013).

Parry, I., R. Williams III and L. Goulder (1999), 'When can carbon abatement policies increase welfare? The fundamental role of distorted factor markets', *Journal of Environmental Economics and Management*, **37**, 52–84.

Peeters, M. (2006), 'Inspection and market-based regulation through emissions trading: the striking reliance on self-monitoring, self-reporting and verification', *Utrecht Law Review*, **2** (1), 177–95.

Peeters, M. and S. Weishaar (2009), 'Exploring uncertainties: the "learning by doing" period has not come to an end, and will still be at the heart of the EU ETS', *Carbon Climate Law Review*, **1**, 88–101.

Perman, R., Y. Ma, J. McGilvray and M. Common (2003), *Natural Resources and Environmental Economics*, Harlow: Longman.

Peterson, S. (2006), 'Efficient abatement in separated carbon markets: a theoretical and quantitative analysis of the EU Emissions Trading Scheme', Kiel Institute for World Economies, Kieler Working Paper 1271, March 2006, available at: http://www.ifw-members.ifw-kiel.de/publications/efficient-abatement-in-separated-carbon-markets-a-theoretical-and-quantitative-analysis-of-the-eu-emissions-trading-scheme-1/kap1271.pdf.

Pindyck, R.S. and D.L. Rubenfield (2001), *Microeconomics*, Upper Saddle River, NJ, USA: Prentice Hall.

Point Carbon (2011a), 'Australia passes emissions trading laws', 8 November 2011, available at: http://www.pointcarbon.com/news/1.1654969?date=20111108&sdtc=1&ref=search (last accessed 6 November 2013).

Point Carbon (2011b), 'China govt think-tank backs 2012 carbon tax: report', 21 November 2011, available at: http://www.pointcarbon.com/news/1.1685822 (last accessed 25 September 2013).

Polinsky, A.M. (1983), *Introduction to Law and Economics*, Boston, MA, USA and Toronto, Canada: Little, Brown & Co.

Prada, M. (2010), La régulation des marchés du CO_2, Rapport de la mission confiée à Michel Prada, Inspecteur Général des Finances Honoraire, 19 April 2010, available at: http://www.ladocumentationfrancaise.fr/var/storage/rapports-publics/104000201/0000.pdf.

PriceWaterhouseCoopers (2009), 'Carbon taxes v carbon trading: pros, cons and the case for a hybrid approach', March 2009, available at:

http://www.timun.net/media/userfiles/file/PWCCarbontaxesandtrading-final-March2009.pdf .

RGGI (2012), 'RGGI 2012 Program Review: Summary of Recommendations to Accompany Model Rule Amendments', available at: http://rggi.org/docs/ProgramReview/_FinalProgramReviewMaterials/Recommendations_Summary.pdf.

Rose-Ackerman, S. (1992), 'Environmental liability law', in T.H. Tietenberg (ed), *Innovation in Environmental Policy, Economic and Legal Aspects of Recent Developments in Environmental Enforcement and Liability*, Cheltenham, UK and Brookfield, VT, USA: Edward Elgar, 223–43.

Rose-Ackerman, S. (1995), *Controlling Environmental Policy: The Limits of Public Law in Germany and the United States*, New Haven, CT, USA: Yale University Press.

Rose-Ackerman, S. (1996), 'Public law versus private law in environmental regulation: European Union proposals in the light of United States and German Experiences', in E. Eide and R. Van den Bergh (eds), *Law and Economics of the Environment*, Oslo, Norway: Juridisk Forlag, 13–39.

Rosen, H.S. (1999), *Public Finance*, Chicago, IL, USA: Irwin.

Roy S. and E. Woerdman (2012), 'End-user emissions trading: what, why, how and when?', in M. Roggenkamp and O. Woolley (eds), *European Energy Law Report IX*, Antwerp, Belgium: Intersentia, 111–42.

Rudolph, S. and T. Kawakatsu (2012), 'Tokyo's greenhouse gas emissions trading scheme: a model for sustainable megacity carbon markets?', Joint Discussion Paper Series in Economics by the Universities of Aachen, Gießen, Göttingen, Kassel, Marburg, Siegen, No. 25-2012, available at: http://www.uni-marburg.de/fb02/makro/forschung/magkspapers/25-2012_rudolph.pdf.

Rudolph, S. and S.-J. Park (2010), 'Lost in translation? The political economy of market-based climate policy in Japan', in C. Dias Soares et al. (eds), *Critical Issues in Environmental Taxation, Vol. VIII*, Oxford: Oxford University Press, 163–84.

Sandbag (2012), 'Turning the tanker: China's changing economic imperatives and its tentative look to emission trading', April 2012, available at: http://www.sandbag.org.uk/site_media/pdfs/reports/Sandbag_Turning_the_Tanker_Final.pdf.

Schwartz, A. and L. Wilde (1979), 'Intervening in markets on the basis of imperfect information: a legal and economic analysis', *University of Pennsylvania Law Review*, **127** (3), 630–82.

Schweizer Eidgenossenschaft (2013), 'Fact sheet – Emission reductions achieved abroad: quality, quantity and carry-over', 28 May 2013, available at: http://www.bafu.admin.ch/emissionshandel/05556/index.html?lang=en&download=NHzLpZeg7t,lnp6I0NTU042l2Z6ln1ad1IZn4Z2qZpnO2Yuq2Z6gpJCGfIR9gWym162epYbg2c_JjKbNoKSn6A–.

Schyns, V. and A. Loske (2008), 'The benefits and feasibility of an ETS based on benchmarks and actual production', in collaboration with FIEC Europe, 27 October 2008 (revised 2 December 2008), available at: http://www.ceps.eu/files/Trilogy_dynBenchmarking_benefits_complete_final021208.pdf.

'Seven pilot provinces and cities explore carbon trading with Chinese characteristics', available at: http://www.sepacec.com/zhxx/xgxx/201210/t20121017_238852.htm.

'Shanghai launches carbon trading pilot and the initial allowances are charged for free', available at: http://sh.eastday.com/m/20120817/u1a6791296.html.

'Shanghai Pudong Development Bank has determined to participate in the carbon trading', available at: http://bank.hexun.com/2012-11-26/148360941.html.

Shavell, S. (1980), 'Strict liability versus negligence', *Journal of Legal Studies*, **9**, 1–25.

Shavell, S. (1984a), 'Liability for harm versus regulation of safety', *Journal of Legal Studies*, **13**, 357–74.

Shavell, S. (1984b), 'A model of the optimal use of liability and safety regulation', *Rand Journal of Economics*, **15** (2), 271–80.

Shavell, S. (1985), 'Criminal law and the optimal use of non-monetary sanctions as a deterrent', *Columbia Law Review*, **85,** 1232–62.

Shavell, S. (1986), 'The judgement proof problem', *International Review of Law and Economics*, **6**, 43–58.

Shavell, S. (1987), *Economic Analysis of Accident Law*, Cambridge, MA, USA: Harvard University Press.

'Shenzhen carbon trading will be launched in June', available at: http://news.xinhuanet.com/politics/2013-04/01/c_124528467.htm. For more information: http://www.cerx.cn/cn/index.aspx.

Sijm, J.P.M. et al. (2004), 'Spillovers of climate policy: an assessment of the incidence of carbon leakage and induced technological change due to CO_2 abatement measures', Energy Research Centre of the Netherlands (ECN), Report ECN-C-05-014, December 2004.

Sijm, J.P.M. et al. (2005), 'CO_2 price dynamics: the implications of EU emissions trading for the price of electricity', ECN, Report ECN-C-05-081, September 2005.

Spier, J. (2011), 'High noon: prevention of climate damage as the primary goal of liability?', in M. Faure and M. Peeters (eds), *Climate Change Liability*, Cheltenham, UK and Northampton, MA, USA: Edward Elgar, 47–51.

Starkey, R. (2012), 'Personal carbon trading: a critical survey', *Ecological Economics*, **73** (C), 7–18.

Stavins, R.N. (1997), 'What can we learn from the grand policy experiment? Positive and normative lessons of the SO_2 allowance trading', *Journal of Economic Perspectives*, **12**, 68–88.

Stigler, G. (1961), 'The economics of information', *Journal of Political Economics*, **69** (3), 213–25.

'The first batch of allowances will be dispatched next month, Guangdong starts to establish carbon emission report system', available at: http://www.ccpit.org/Contents/Channel_50/2013/0226/356516/content_356516.htm.

'The pilot carbon trading is undergoing the last stage: how to design allowances becomes the focus', available at: http://szsb.sznews.com/html/2013-04/04/content_2431510.htm.

Tietenberg, T. (2000), *Environmental and National Resource Economics*, Reading, MA, USA: Addison-Wesley.

Tietenberg, T., M. Grubb, A. Michaelowa, B. Swift and Z.X. Zhang (1999), 'International rules for greenhouse gas emissions trading: defining the principles, modalities, rules and guidelines for verification, reporting and accountability', UNCTAD/GDS/GFSB/Misc.6, United Nations Conference on Trade and Development (UNCTAD).

Tokyo Metropolitan Government (2010), 'Tokyo cap-and-trade program – Japan's first mandatory emissions trading scheme', March 2010, available at: http://www.kankyo.metro.tokyo.jp/en/attachement/Tokyo-cap_and_trade_program-march_2010_TMG.pdf.

Trotignon, R. and A. Delbosc (2008), 'Allowance trading patterns during the EU ETS trial period: What does the CITL reveal?', Caisse des Dépôts, Climate Report No. 13, June 2008.

Tuerk, A. et al. (2009), 'Linking of emissions trading schemes: Synthesis report', Climate Strategies, 20 May 2009, available at: http://www.climatestrategies.org/research/our-reports/category/33/148.html.

Tuerk, A., M. Mehling, C. Flachsland and W. Sterk (2009), 'Linking carbon markets: concepts, case studies and pathways', *Climate Policy*, **9** (4), 341–57.

Turner, R.K., D. Pearce and I. Bateman (1994), *Environmental Economics: An Elementary Introduction*, New York, NY, USA: Harvester Wheatsheaf.

Union of Concerned Scientists (2007), 'Existing cap-and-trade programs to cut global warming emissions', available at: http://www.ucsusa.org/global_warming/solutions/big_picture_solutions/regional-cap-and-trade.html.

Universal Declaration of Human Rights, UNGA Resolution 217A (III), UN Doc A/810 (1948) 71, available at: http://www.un.org/en/documents/udhr.

University of Copenhagen (2009), 'Synthesis Report – Climate change: global risks, challenges and decisions', available at: http://climatecongress.ku.dk/pdf/synthesisreport (last accessed 19 May 2013).

Van Dijk (2011), 'Civil liability for global warming in the Netherlands', in M. Faure, and M. Peeters (eds), *Climate Change Liability*, Cheltenham, UK and Northampton, MA, USA: Edward Elgar, 206–26.

Van Renssen, S. (2012), 'The fate of the EU carbon market hangs in the balance', *European Energy Review*, 12 April 2012, 1–8.

Varian, H.R. (2003), *Intermediate Microeconomics: A Modern Approach*, 6th edn, New York: W.W. Norton & Company, Inc.

Vickrey, W. (1961), 'Counter-speculation, auctions, and competitive sealed tenders', *Journal of Finance*, **16** (1), 8–37.

Weber, R. (1997), 'Making more from less: strategic demand reduction in the FCC spectrum auctions', *Journal of Economics & Management Strategy*, **6** (3), 529–48.

Weishaar, S. (2007a), 'CO_2 emission allowance allocation mechanisms, allocative efficiency and the environment: a static and dynamic perspective', *European Journal of Law and Economics*, **24** (1), 29–70.

Weishaar, S. (2007b), 'The European CO_2 emissions trading system and state aid: an assessment of the grandfathering allocation method and the performance standard rate system', *European Competition Law Review*, **28** (6), 371–81.

Weishaar, S. (2008a), 'The European emissions trading system: auctions and their challenges', in M. Faure and M. Peeters (eds), *Climate Change and European Emissions Trading: Lessons for Theory and Practice*, Cheltenham, UK and Northampton, MA, USA: Edward Elgar, 343–62.

Weishaar, S. (2008b), 'Ex-Post-Korrektur im Europäischen CO_2-Emissionshandel: Auswirkungen der Rechtsprechung für Deutschland', *Zeitschrift für Europäisches Umwelt- und Planungsrecht*, Vol. 3, pp. 148–51.

Weishaar, S. (2009), 'Towards auctioning: the transformation of the European greenhouse gas emission trading system – present and future challenges to competition law', Alphen aan den Rijn, The Netherlands: Kluwer Law International.

Weishaar, S. (2012), 'EU state aid law and national climate regulation', in M. Peeters, M. Stallworthy and J. de Cendra de Larragán (eds), *Climate Law in EU Member States: Towards National Legislation for Climate Protection*, Cheltenham, UK and Northampton, MA, USA: Edward Elgar, 89–109.

Weishaar, S. and F.G. Tiche (2013), 'Hybrid emissions trading systems: what about efficiency?', in M. Faure and G. Xu (eds), *Using Economics to Improve Regulation: The Case of China*, Abingdon, UK: Routledge, 221–39.

Weishaar, S. and E. Woerdman (2012), 'Auctioning EU ETS allowances: an assessment of market manipulation from the perspective of law and economics', *Climate Law*, **3** (3), 247–63.

Western Climate Initiative (2010), 'Design for the WCI Regional Program', available at: http://www.westernclimateinitiative.org/component/remository/general/program-design/Design-for-the-WCI-Regional-Program.

Williams, A. (2007), 'The European Union emissions trading system: a cement industry perspective', presentation delivered at the North America and the Carbon Markets Conference, Washington, DC, 17–18 January 2007.

Wittman, D. (1980), 'First come, first served: an economic analysis of "coming to nuisance"', *Journal of Legal Studies*, **9** (3), 557–68.

Woerdman, E. (2004), *The Institutional Economics of Market-Based Climate Policy*, Amsterdam: Elsevier.

Woerdman, E. (2005), 'Tradable emission rights', in J.G. Backhaus (ed), *Elgar Companion to Law and Economics*, Cheltenham, UK and Northampton, MA, USA: Edward Elgar, 364–80.

Woerdman, E. (2013), 'Lobbying in the European Union emissions trading scheme: inefficiencies caused by industry rent-seeking', March 2013, University of Groningen, Working Paper Series in Law and Economics.

Woerdman, E. and J.W. Bolderdijk (2010), 'Emissions trading for households: a behavioral law and economics perspective', University of Groningen, Working Paper Series in Law and Economics.

Woerdman, E., O. Couwenberg and A. Nentjes (2009), 'Energy prices and emissions trading: windfall profits from grandfathering?', *European Journal of Law and Economics*, **28**, 185–202.

Woerdman, E., S. Clò and A. Arcuri (2008), 'European emissions trading and the polluter-pays principle: assessing grandfathering and over-allocation', in M. Faure and M. Peeters (eds), *Climate Change and European Emissions Trading: Lessons for Theory and Practice*, Cheltenham, UK and Northampton, MA, USA: Edward Elgar, 128–50.

Zaklan, A. (2013), 'Why do emitters trade carbon permits? Firm-level evidence from the European emission trading scheme', EUI Working Papers, RSCAS 2013/19.

Zwingmann, K. (2007), 'Ökonomische Analyse der EU-Emissionshandelsreichtlinie, Bedeutung und Funktionsweisen der Primärallokation von Zertifikaten' ('Economic analysis of the EU

Emissions Trading Directive, significance and working of the primary allocation of certificates') in der Reihe *Ökonomische Analyse des Rechts* (in the series *Economic Analysis of Law*), Wiesbaden: DUV Gabler Edition Wissenschaft.

Index